Radical Markets in Taiwan: Extended Reading with Local Perspectives

JU-CHUN KO, Assistant Professor of
National Taipei University of Technology

Copyright © 2020 Dr. J. C. KO,
Assistant Professor of National Taipei University of Technology
All rights reserved.
ISBN: 9798590434688

DEDICATION

For my beloved Taipei, and Taipei Tech.

Radical Markets in Taiwan:
Extended Reading with Local Perspectives

Radical Markets was selected as ***Books.com.tw*** Book of the Year 2020, and the Best Business and Economist Book of the Year by the Economist. The authors were both on Bloomberg's list of 50 Most Influential People. At the invitation of a publisher in Taiwan, the following three of all chapters were especially written as extended reading hopefully to enlighten local readers in Taiwan. The other two chapters that have not been published are compiled here, and four episodes of BBFT Podcast with related topic about Radical Markets. One bonus chapter from Tim O'Reilly's book *WTF? What's the Future and Why It's Up to Us.*

CONTENTS

	Acknowledgments	i
1	Preface: *Radical Markets* in Taiwan	1
2	Houses here look a bit old	4
3	Green card lottery meets diplomatic talents	9
4	Hand grappling techniques of the Monopolizing Octopus	15
5	Epilogue: Extended reading with local perspectives	20
6	Follow Columbus in the Moments of History - A Foreword for Tim O'Reilly's new book *WTF? What's the Future and Why It's Up to Us*	24
7	BBFT Podcast #EP2 *"A meeting in the forest to decide the future of human finance!"* ft. Dr. Tom Lam	29
8	BBFT Podcast #EP5 *"The University of Chicago genius proposed five moves to radicalize the market to solve the capital problem!"* Solo	49
9	BBFT Podcast #EP10 *"Someone slipped the phone and became the unemployment crisis for everyone after ten years!"* ft. Dr. Suji Yan	60
10	BBFT Podcast #EP23 *"Liquid democracy, how much value we hold?"* ft. Kin Ko, LikeCoin Chain	87
11	BBFT Podcast #EP44 *"The post-epidemic world is in the palm of your hand!"* ft. Audrey Tang	110

ACKNOWLEDGMENTS

Department of Interaction Design,
National Taipei University of Technology.

1. PREFACE
RADICAL MARKETS IN TAIWAN

I recall a summer afternoon when I was in a renowned Taiwanese entrepreneur's office chatting over the immediate economic outlook that the world was about to face. It was a time before the deadly pneumonia outbreak, nor any need for home quarantine or WFH (working from home). What we saw was the revolutionary possibilities for the whole humanity brought by virtual reality, new media technologies, blockchain, Cryptocurrency and artificial intelligence. Looking outside the window, the entrepreneur told me about the income inequality that he witnessed recently. A while back, he went on a fund raising trip where he met all the rich business tycoons that any ordinary person may not get to meet in all three lifetimes. They were from Taiwan, China, Hong Kong, the US and Europe. He saw a three-story tall aquarium specifically for keeping two red dragon fish; a super sumptuous banquet where he downed a 2-millions-per-bottle red wine. As he talked, the entrepreneur rested his eyes on the distant mountains outside the window and said, "Look at the mountain there. Why doesn't it belong to us? Why are the mining rights of certain diamond mines and gold mines on the other side of the earth, despite being the property of all earth (citizens), inexplicably authorized by specific nations to private companies which would in turn impact on the global economy? Further, why are these assets and resources that should have been shared by all earth citizens be assigned, without asking, to some specific groups of people, be they nations, governments or consortia."

To a large extent, private properties turned public possessions somehow entail that you have a share in everything owned by other people around you.

Right. Governments are supposed to allocate resources on our behalf and benefit us. But why is it that what we see is governments taking our taxes to control banks by holding their shares, and the banks then loan out the money to consortia which use the money for further investments whose profits will ultimately return to the banks and governments, and on and on the transactions continue. Based on the concept of "Equality for all", the regular middle class should have been able to benefit from establishing social resources and maintaining political borders. But now it looks like only a small handful of people are always getting the profits. What is going wrong with the society?

This entrepreneur was known for his shrewd commentaries of current affairs, and many of his fans often had their innermost justice-related sentiments stirred up by the views that he published. Yet, the entrepreneur was an entrepreneur after all; he was no capitalist, so his opinions remained as opinions with little academic impact. When confronted, those in charge of power and resources could always nonchalantly claim that "Such are the rules, and I can only follow them," or "In a capitalist market, freedom reigns supreme." And that was that. Now that the book, *Radical Markets*, has been published, all the above would be given a new perspective. The book was co-written by Glen Weyl, the youngest ever Economics professor in the University of Chicago, and Eric Posner, the most distinguished member in the best-known family of legal scholars in the west. Vitalik Buterin, a tech genius in blockchain, instantly became a fan and volunteered to write a foreword for the book upon its publication and even gave a shout-out to the *Radical Markets* camp, "Whatever practical tech support you need, I am there," offering all his contacts and prestige in the tech circle and in the blockchain sector.

In *Radical Markets*, the two authors walk away from what is deemed by many as the alter of high achievers; like any social scientists that care about the current state of the world, they resort to mathematics, economics, social science and legal studies, as well as their understanding and imagination for technologies in an attempt to raise a warning against the contemporary capitalist economy that is walking on a tight rope. No, it's more than a warning. The two authors have shown their absolute courage in going against the tide to tell the truth and propose five fundamental remedies. They are like Gandalf and Frodo in our contemporary society where monopoly and totalitarianism are being carried out in the name of market competition.

As the social economics has developed in the past few decades, books like The Wealth of Nations by Adam Smith, A Treatise on Money by Keynes and Hayek's views on free market have all become the Rings of Power that entitle the ring owners, i.e. the capitalism practitioners, to speak with authority on topics like markets, capital and freedom. In reality, the capitalists have been carrying out one after another totalitarian scheme. Their so-called "Invisible Hands" that are meant to be non-human hands but hands of divine intervention, and supposed to be part of the free market mechanism, have now become human interference playing God, trying to be invisible in their unjust dealings.

"The rich...are led by an invisible hand to make nearly the same distribution of the necessaries of life, which would have been made, had the earth been divided into equal portions among all its inhabitants, and thus without intending it, without knowing it, advance the interest of the society."

"Data, one of the most valuable commodities in the digital economy, are collected and monetized by companies such as Google and Facebook, but the users who create these data receive no direct compensation. A much-needed market in data simply does not exist."

The two authors of *Radical Markets* risk being misunderstood and scorned by the masses to unveil the reality behind the Rings. How we can follow them across high mountains to find the solution that may or may not exist depends on whether or not we take action after reading the book.

"Even if we don't sell you on all our ideas, we hope this book will open your mind to a new way of imagining the economy and politics. This challenging moment, when long-held assumptions are being overturned, is ripe for radical rethinking."

I was invited by the publisher of the book's traditional Chinese translation edition to write a response to Chapter 1, 3 and 4 from a local perspective to serve as extended reading for readers in Taiwan and those who can read traditional Chinese. The English translation of this book is intended for more readers to know Taiwan's perspectives on some of the innovative ideas in the book and their connection with Taiwan.

2. HOUSE HERE LOOK A BIT OLD

I recall a few years back when I was asked by a friend to host his friends from Shanghai on their first visit to Taiwan. We agreed to meet near the MRT Dongmen Station. We went through the alleyways. I as the host showed them the famous landmarks on Yong Kang Street and planned on taking them to sample soup dumplings in Din Tai Fung. All along the way, they showed great interest in the street scenes. I noticed one person was particularly curious about the buildings on both sides of Yong Kang Street. She was on the verge of saying something but then held back her words. I gathered whatever was on her mind might remain private or unresolved for a bit, so I went on to talk about the history of Yong Kong Street. I began with the Qing Dynasty when the area was a wilderness, went onto the construction of the Liugongjun Canal, the Bao Gong Cinema and the opening of the Da An Forest Park. This visitor from Shanghai finally broke her silence, looked at a row of under-five-story old houses on Yong Kang Street and said in her perfect Shanghai accent, **"How come your houses here all look a bit old?"**

It happened to be a time when the Wenlin Yuan Urban Renewal Dispute occurred and housing justice was being widely discussed; my friends with academic studies on public administration were also commenting on the various regulations of urban renewal. This visitor is a friend of a friend who went on my podcast as a guest before. On hearing her words, "Houses here look a bit old," I was speechless, not able to respond. I held my thoughts on quasi-democratic values and the price of totalitarianism, without uttering a word.

Notwithstanding this Shanghai visitor's words carried some undercurrents of misconception and even a bit of malicious mockery and attack. Now when I pass through certain streets that look distinctly out of place among its neighborhood, for instance, several sections on the Civic Boulevard or the corners around Qiang Shu High School behind the MRT Guting Station or Wenshan District and Nangang District as pointed out by many people on the social media, I can't help but think: wouldn't it be fantastic if there were an effective approach to urban renewal that could also protect the rights and interests of existing residents?

A few years on, I've become an assistant professor in National Taipei University of Technology (Taipei Tech). I am passing through Renai Road more often than ever where most of its sections have been turned into rich neighborhoods with luxurious houses. Yet, I came to notice one of the houses. Despite its ostensibly grand, beautiful and lavish exterior, the house was always locked up with leaves and cardboard boxes scattered all over the place. I could not possibly imagine how an opulent building like this on Renai Road would remain uninhabited. Had there been a case of Rose n' Siren Eyes or Taiwanese Paranormal Encounters occurring inside the house? I noted the address: No. 72, Sec. 2, Renai Road, and looked it up on internet to discover that it was the legendary luxury home which its housing developer built 16 years ago but was unable to sell. There were (and still are) so many people out there with no place to go to on a cold miserable night while this being one of the first grand houses on Renai Road had somehow remained unoccupied for 16 years. And that was only one of such examples. If you ever visit Taipei's Xinyi District, you will surely see this famous house standing as part of the Tao Zhu Yin Yuan luxury apartments dubbed "Spinning Luxury Homes". After going on the market for a few months, its first transaction was reportedly made with a subsidiary under the same construction company, hence the widespread speculation that the house was sold as an insider trading.

Evidently, monopolizing assets and leaving them idle are a waste to the entire society. The proposal of Luxury Tax, Empty Houses Tax and Housing Tax for Over-Accumulation are applauded by many people, while some others find it unnecessary and that it will interfere with people's rights to freely possess and control assets in a capitalist society, hence such mentality as the following: "What is wrong with me wanting to leave the house unoccupied after buying it?" or "What's wrong with wanting to buy an old metal house in the capital and live in it for 100 years?"

From the perspectives of *Radical Markets*, resource misallocation and immobility has to be resolved. With rational analysis, Glen Weyl as an economist states clearly that many ineffectually incurred taxes that are intended to solve all kinds of injustice are all missing the mark, and more strictly speaking, an unnecessary waste.

Is there a kind of tax system which, once announced, will save all the time spent on tax discussion and solve these problems in one go: resource idleness & waste and a small handful of people monopolizing the assets and refusing to comply with the bigger need for social environments and development and undertake adjustments accordingly? Housing justice is only one aspect, others including transportation development (with the case of Hyperloop, a high-speed metropolitan network, being cited by the authors as an example). Is there a tax system that can solve all the problems?

Indeed, the authors propose such a Hollywood-grade innovative tax system called the Common Ownership Self-Assessed Tax (COST). The so-called self-assessment means you can determine the value of an item yourself and pay the tax accordingly. If you consider your house worthy of $1, then you pay one dollar's worth of asset tax. Or you can define your house as worthy as $10 billion (so that you appear to be rich) and will by contrast pay the amount of tax corresponding to the $10 billion asset (the book citing 7% as the ideal COST tax, so that amounts to $0.7 billion in tax per year). The common ownership means that anyone can go through the auction mechanism (supplemented by some necessary resources/assets) as a protection mechanism to auction in (acquire) the resources/assets that you have put a price on. That is to say, you have lost the right to perpetual ownership of most commodities, and all the resources and assets that have an impact on the social development can be traded anytime and remain fluid, as is secured by compulsory law enforcement.

On reading this, many people, I believe, would exclaim: Isn't this communism? Isn't this totalitarianism?

Relax. Yes, to some degree, the authors' COST proposal is indeed made in the spirit of Communism or Socialism, but essentially, what lies behind the proposal is not a farther-reaching edition of Totalitarianism but an integration of extreme free market transactions in a bid to resolve many of the aforementioned resource idleness/misallocation impasse. But is their viewpoint really feasible, i.e. "no longer owning most of the things". Many criticize the authors' proposal as Socialist or Communist. But truth be told, we do live in a time when we can accept AirBnB as a co-shared housing solution, and when Japan has launched a new housing approach, i.e. HafH, offering "Fixed price all-you-can-stay" service, whereby monthly payments of a certain amount will give you unlimited access to many hundreds of studios and houses around the world. Similar trends include co-shared spaces and scooters, for instance, WEMO and GoShare in Taiwan. If we can accept having no permanent ownership of any item, then why can't we extend such concept/design of asset ownership to cover almost all items?

The authors reference the god of auction, Nobel Laureate in economics, William Vickrey, and call such assets covered by common ownership as Vickrey Commons. Mind you, this is not about co-sharing properties, but about being able to auction off the ownership of common properties at any time. To a larger extent, the ownership can be changed to a common entitlement. Alternatively, it's not a common ownership but a mechanism where its ownership can be possessed, governed, entitled and traded at any time. Such an extreme market viewpoint can be applied to the shared ownership of human resources, i.e., the COST of human capital.

"The talented enjoy a kind of freedom, as they can select from among a variety of appealing jobs. These jobs allow them to quickly accumulate capital that they can depend on, as they age, if they do not like the jobs that are available, or pick and choose among different levels of labor (part-time, enjoyable or low-paying jobs in the nonprofit sector, etc.). Those with fewer marketable skills are given a stark choice: undergo harsh labor conditions for low pay, starve, or submit to the many indignities of life on welfare. Yet the waste of social resources when a talented person fails to realize her potential is far greater, and arguably their failure to work should be punished more harshly.

A COST on human capital would ameliorate this form of unequal freedom by requiring the talented people to pay a tax if they do not want to work in a job that is most efficient for society."

To me, the above is one of the most shocking paragraphs in the book.

Are we able to accept such radical ideas? I often hear my friends talking about at what age they would like to start an early retirement. They make it sound as if they work hard enough, save enough capital by a certain age, they will be entitled to a life of no more work but play. Yet, the *Radical Markets* authors being the high-achievers stand with us on this matter in the sense that they resort to their economics expertise and propose that the injustice of such thinking, i.e. early retirement upon having accumulated enough personal wealth, desperately needs to be brought to our attention and debated over. After reading Chapter 1, I happened upon a translated article on BBC Chinese website published in 2018, A British Woman's Confession: Poverty opened my eyes to money. Excerpted from the article is as below:

> "In the university dormitory, no matter where we came from, we all look more or less the same from the exterior. But during holidays, these new friends of mine would all disappear. They have all either gone home or taken on internship in London. Internship sounds great but it's not a paying position. I have no money to go on (non-paying) internship, so I worked in a local shoe shop selling shoes that I myself cannot afford. In so doing, I can pay for my rent and carry on living in the university dormitory... In our first class meeting, I felt what I knew fell far behind the guy that had done 10 intern jobs."

This instantly reminds me of my friends living in luxury homes who went to study in New York during school holidays while they were still in senior high school or university; they took up part-time jobs with McKinsey, while I worked as an intern for NT$50 a day producing online advertisements in a company located in a residential building in the Zhonghe Industrial Zone. What's more startling is that originally I was only envious/jealous of my friends living in luxury homes, but little did I know that being competent enough to be earn a wage of NT%50 a day was already deemed a triumph. There were people that could not get intern jobs and thus lost the opportunity to surpass or get equal with their peers. Based on this recollection and reflection, plus the following astounding data quoted by the authors:

> "...we see that the share of income taken by the top 1% of earners has roughly doubled from its trough of 8% in the mid-1970s to its recent peak of 16%...There has been a nearly 10% drop over this same period in the share of national income in the United States..."

I immediately became a COST fan. Should the COST be implemented one day, can we accept its ensuing ripples such as lowering the bubble-like housing prices and commodity prices to avoid the 7% annual ownership taxation resulting in the capital value of everyone's private possessions such as notebook computers, houses, stock shares and savings being reduced by half in return for a regular monthly allowance from the government, i.e. a few tens of thousands of NT dollars per month (insistently derived from the 7% COST paid by the rich). What is our stance in this? That would be another matter.

Radical Markets: Uprooting Capitalism and Democracy for a Just Society
By E. Glen Weyl, Eric A. Posner.,
Chinese Translation by Zhou YiFang
Chinese Edition published by Gusa Publishing on 2020/04/29
Extended reading for Chapter 1 of the book

3. WHEN GREEN CARD LOTTERY MEETS DIPLOMATIC TALENTS

A Hong Kong friend of mine is a wizard of games. He started his own mobile games company 20 years ago. There was no Java Virtual Machine (JVM) back then, let alone iOS or Android. His company made a big fortune after a few years and continuously received investment from several renowned global venture capital firms. He had fantastic fun with his business and was a well-known figure in the industry of games. A while ago, he pulled out and went into a brand new market to begin another venture: Blockchain Likecoin and Liker Land. He ran his new venture spectacularly for a few years attracting over 10,000 users, half of whom were based in Taiwan. Taiwan's users were very fond of his innovative services which had successfully prompted Taiwan's creators to develop many new original content on his platform. Some creators' ecologies had become more streamlined and well circulated thanks to the bloackchain services that he developed. This Hong Kong friend of mine had high hopes over Taiwan's free and mature market that is willing to accept new things. He hoped to have his new company based in Taiwan for long-term development.

In the beginning, this friend successfully joined a famous startup acceleration program in Taiwan; later, he put together some of his previous business investment records to secure an entrepreneur visa in Taiwan. One year later, due to the nature of his company, he could no longer renew his visa to stay in Taiwan. He once asked me: Is there any visa application route via collaboration or recommendation so that he could contribute his strengths while staying here in Taiwan on a more flexible and long-term basis? We had many joint projects, often got together and chatted over anything under the sun. Naturally I would very much like him to stay in Taiwan. In the next few months, I enquired over many official routes and ways for him to stay. He himself made a lot of efforts too. Due to various reasons, he didn't get to stay in the end. This spring, he left Taiwan, a place that he took very seriously and was very fond of, and returned to Hong Kong to carry on with his endeavors. On his next visit to Taiwan, he'd be going through a different immigration gate that would impose more restrictions. It got me thinking that there must be quite a lot of outstanding foreign workers in Taiwan like this friend of mine. If there was a mechanism whereby I could submit to the government a recommendation letter based on my expertise in the industry or in the academia in favor of his stay in Taiwan, I was certainly willing to do so.

I have another friend, who was born and raised in Taiwan and now works in the US. After graduating from the Electrical Engineering Department of National Tsing Hua University, he went on to study in the US and later secured a visa to stay and work. Now he serves in a very successful AI startup. Every now and again, he'd try to persuade me to submit applications to teach in the US, and say that he'd write recommendation letters for me. Besides passing me information about universities and immigration law firms, he would periodically send me pictures of him driving his roadster at 5pm after work to sunbathe on a beach near his company in San Diego. That was what he did throughout the week. As I still have projects that I want to do in Taiwan, I have not got round to those emigration strategies he sent me. One day, he sent a message saying that his sister was migrating to the US. The sister went through a very uniquely peculiar route, not through the O1 or V1 plan but through the lottery. For over 10 years, this friend had been routinely and punctually visiting the online immigration lottery website on behalf of his sister and continued to do so even after he himself had secured a job. His sister got lucky with the lottery, and a few months later, he told me about it. He said with excitement that his sister was soon to move to the US and would be staying with him in San Diego before getting a job.

On hearing this, I was quite surprised. I had long heard of this green card lottery thing since I was a child. I always thought it was an internet scam. Little did I know it was all legit. The friend said that of course it was legit. Its official name is "Diversity Visa (DV) Program", designed to increase diversity in the U.S. immigrant population offering 50,000 visas per year. As this is a very popular immigration route, what with limited availability and vast numbers of applicants, there are numerous online scams trying to con people by claiming it's easier to be awarded such visa with fabricated applications. The lottery program itself is legitimate, but there are many online fraudulent versions of it. Two years later, over a meal with this friend, he sighed and said that his sister had decided to come back to Taiwan for no other reason than that she could not find a job in the US (as her entry was granted by lottery and she did not have a US degree) and that she was not used to the American culture, food and environments. After a big fight with the brother, she decided to settle in Taiwan and find a job here.

On hearing this story for the first time, my reaction was: what a pity to give up such a rare opportunity of having won the lottery. But then, after much thought, the sister's decision came as no surprise at all. It was not her own intention to draw the lottery, and she only knew one person in the US, i.e. her brother, with all her other family members and friends in Taiwan, so it was only normal not able to fit in. And yet this means this friend's sister had become a case of low-efficiency or zero-efficiency talent mobility. Here we are in 2020, and the United States of America that has been well reputed for its advanced data management and national strategic thinking should still resort to annual green card/visa lottery to balance the immigration population. Is there not a better and more efficient way of balancing domestic or international talent or population migration? That was my question at the time.

In *Radical Markets*, the two authors devoted a lot of space to the exploration of international talents, population and labor migration. In this chapter, they first point out a very important observation and research result, namely, the gap between rich and poor, which is happening not only domestically but also worsening between countries in recent years.

"...persistent differences in mass living standards across countries were unknown until the late nineteenth century. Even the most extreme gaps, such as between China and the United Kingdom, were only a factor of 3. This contrasts with the 10 to 1 gap that opened up by the 1950s."

"Inequality across countries increased from about 7% in 1820 to about 70% in 1980...Together these patterns imply that inequality across countries has gone from a relatively insignificant phenomenon in the grand scheme of global inequality, (accounting for only a little more than 10% of global inequality in the 1820s), to being the dominant source of global inequality, (accounting for two-thirds or more in the second half of the twentieth century and still today accounting for 60–70% depending on whose measurements you rely upon)."

So here is their proposal: if there can be an increased quota of talent mobility in rich countries, and competent workers in poor countries can easily move to and work in a more socially advanced country, this would enable *"...roughly a 20 percent increase in global income"* according to the authors' calculation.

No wonder their views are considered radical in the radical market-ism: what exactly can be done to enhance talent mobility? The two authors propose a radical chapter under the headline, "Uniting the World's Workers" plus an equally radical solution, Visas Between Individuals Program (VIP).

What is radical about the proposal? The authors propose the solution to talent mobility is by uniting the world's workers as is spelt out in the headline. Those with some knowledge of social science would perhaps immediately pick up something odd here. Doesn't the above headline sound somewhat familiar? That's right. A second look would remind one of the Communist slogan that took the world by storm back in the days: "Workers of the world, unite!" The Communism allusion may be missing in the Chinese translation but the English text, "Uniting the World's Workers", is a syntactical rearrangement from the original Communist slogan. No wonder this chapter and its content have attracted much criticism against the two authors. Many readers of the free market faction, on reading this sentence, would inevitably let out a scornful remark, "Communists making a comeback?" Once they read the content, they will probably faint out of shock (or anger).

"Anthony learns of a new program by the State Department that allows him to sponsor a migrant worker. But what's in it for him? Unlike Google, he can't simply place the worker in an office and expect him or her to generate revenue for him.

... the two agree that Bishal will work for Anthony for one year in the United States...Anthony has to use his savings to buy Bishal a flight ticket. They agree that Bishal will reside in Anthony's spare room...If Bishal disappeared, Anthony would also be fined. We don't think the fine should be steep but should be enough to hurt."

Isn't this a renewed version of servitude? Not only will Communism make a comeback but slavery is also to return?

To be honest, this is the chapter which I found most dubious when reading the book. Besides appearing to be highly idealized, it seems to have leaned too much towards data calculation and strategy deployment to have overlooked our accommodation and respect for equality. Reading this chapter elicits the questions: What is happiness? What is an ideal nation? Does an increase of national income equate happiness? For the sake of a narrowly defined happiness, does every worker in not-so-rich countries genuinely desire to work as a migrant worker in Europe, the US and Japan?

> *"While such an outcome is far from true equality, it is probably the best that can be hoped for in the near term."*

The authors confess near the end of the chapter that they are very clear that the content discussed in this chapter is quite far off the society's current or ideal conception of equality. But the authors' ambition is not merely about exploring the highly skilled talent mobility strategies in various countries in recent years but expanding to examine the low-wage low-rung manual laborers that have already migrated to advanced rich countries. That's why the authors propose their view as necessary despite its temporary shortcomings. They authors think that regardless of whether we go ahead with their proposal, i.e. the VIP, the low-wage labor predicament has already been existent. It's only that the rich class who have a say in such policies in wealthy countries would rather turn a blind eye regarding such matters. The result is that such issues pertaining to labor induction and management are tucked away in a small number of bottomless evil foreign labor rings and routes. These low-wage foreign labor are often bereft of security and end up doing more and more high-risk jobs. Yet there are no appropriate management and combined measures in place to tackle the relevant problems. Instead of leaving the management to remain out of sync, it's better to implement more systematic, mutually-beneficial measures so that those with discernibility can act on such incentives as bonuses or remunerations and bring into their own country workers that are more productive and better protected. This is a stone that kills two birds in the sense that wealthy countries can have influx of better and more talents into all levels of employment, and workers in developing countries will have more opportunities to work around the world as one unified job market in return for an income that they can send back to their home country to contribute to its growth and bridge the income inequality between wealthy and poor countries.

Here are the ultimate questions: 1. Will the VIP attract the intellectuals with discernibility or will it only appeal to the likewise low-wage native workers who are in it for some passive income and indiscriminately bring in incompetent workers? After all,

> *"Our aim is to involve working-class people who would be attracted by the financial benefits of sponsorship. A low-income person who could net $6,000 from sponsoring a low-skilled migrant worker would significantly increase their well-being; in contrast, a middle-class or wealthy person is not likely to find such an opportunity attractive."*

And 2. What incentives are available for wealthy countries to open up borders to even more workers from developing countries than before? Is the VIP system overly idealistic in a social climate dominated by conservatives?

Regarding both questions, the authors first mention the J-1 visa program that is being implemented in the US to prove that the sponsoring system will be popular and effectively managed.

"While the J-1 program was initially designed for cultural exchange, Congress has permitted its use for what is essentially low-wage nanny work... While some people argue that the au pairs are exploited, we have not found any rigorous studies that document abuses."

"They rely on intermediary institutions (private companies) to match American sponsors and foreign workers, train house helps and follow up on workers' employment and housing situations upon their arrival in the US, which is all regulated and managed by the State Department. (Note: This approach is similar to Taiwan's system on migrant workers and house helps.)"

The epilogue in the book carries a supplement that clears my doubt:

"Further, suppose wealthy countries and poor countries reach an agreement where the new VIP system will replace today's practice of using relief fund as a form of assistance to poor countries, and the VIP dictates that all countries agree to share part of the COST income. Given such a premise, wealthy countries can supply relief fund to poor countries at any point in time. However, if poor countries become richer as they develop, such transferred payment will stretch out and the bilateral payment will become equalized. As such, this would give citizens of rich countries an incentive to develop poor countries, as well as giving citizens of poor countries a reason not to resent too much the prosperity of wealthy countries. Together these two features would help tilt the scales of opinion in wealthy countries in favor of opening migration further to aid the development of poor countries."

I could finally understand why it'd be a radical and pragmatic solution if we can replace the Dollar Diplomacy that has been much criticized recently with Talent Diplomacy. If we see education as the foundation of a powerful country, then talent mobility would be the backbone of a country's vibrant economy (the early development in the US refers). One can even say that a further transparent and free talent mobility with reinforced incentives will become the catalyst for a global unified economy and go on to promote a balanced and sustainable economic development for all countries, despite some slightly "ruffled sensibilities". ***"This is a moral gain relative to the hypocrisy of our current system and perhaps the only plausible way toward a more just international order."***

Radical Markets: Uprooting Capitalism and Democracy for a Just Society
By E. Glen Weyl, Eric A. Posner.,
Chinese Translation by Zhou YiFang
Chinese Edition published by Gusa Publishing on 2020/04/29
Extended reading for Chapter 3 of the book

4. HAND GRAPPLING TECHNIQUES OF THE MONOPOLIZING OCTOPUS

I recall that in my youth I used to go to great lengths just to buy computer parts in Taipei's famous Guanghua Digital Plaza; I wanted my self-assembled computer to run faster so that it can download more software for new games. There was no online malls back then. So it was the heyday for the Guanghua Digital Plaza. A short Bade Road was lined with many dozens of 3C shops, with a few more dozens of them in its basement, the "International Electronics Plaza". For the sake of saving a few NT dollars, I'd go through all the shops looking for the best bargain, asking the shop owner, "How much is your 7200 RPM?" or "How much is your 1333 RAM? Is it in stock?" The shop owners were fed up with all these questions and started to put on densely arranged lists of price quotes which were updated daily. When new products were launched, crowds of computer users would swarm outside the shops peering at the lists with much the same emotional eagerness and intensity of students trying to elbow for a view of the result of the university entrance exam. I'd stretch my neck to see if my desired item was on the list, took notes, and compared prices. Sometimes I had to visit over 10 shops just to find a bargain that saved me NT$100, and I'd be over the moon when I did. Sometimes, after a whole afternoon, I only managed to save less than NT$50, and worse still, when I returned to the shop for that cheaper-by-NT$50 bargain, that item might have been scooped by the person right before me. Imagine walking for 10 minutes to get back to the first shop only to be told, "Sorry. Out of stock." That was how I spent my winter and summer holidays enquiring prices, assembling, dissembling and re-assembling the computer.

This went on till one day when I accidentally read a report on a magazine that struck me like thunder, burning me like charcoal, banishing all my remaining youthful innocence. Written in bold type in the magazine were words such as "Many shops in Guanhua Digital Plaza actually belong to the same boss as part of a chain…" Another report stated, "Consumers would previously go to Guanhua Digital Plaza to enquire and compare prices, when actually many stores are under the same ownership, so the prices have always been controlled by the sellers. Such asymmetric information has earned immense profits for some business proprietors…" So when I was zipping in and out of shops burdened by my sachet of school books hunting for the best deal, those "best shops" with "honest pricing" actually did not exist. Those shops very likely belong to the same person or group of persons. These people open up a bunch of shops to sell identical goods. Since customers like to compare prices, out of necessity and out of habit. "Love comparing prices, huh? OK you can have a field day. Some shops would be NT$10 cheaper or pricier. No matter which shop you go to, you'd still fork out the money and it all goes into my pocket." Such practice has been around till the emergence of online shopping.

Here is another example. I believe many readers still remember the panic buying of toilet paper a while back. As recorded in Wikipedia: "Taiwan's panic buying of toilet paper in 2018" dubbed as "Toilet Paper Chaos" or "Defecation Chaos" refers to the phenomenon of panic buying of toilet paper following speculation of price increases caused by manufacturers' indecent marketing techniques and media coverage. Back then, the Fair Trade Commission did eventually launch an investigation on toilet paper manufacturers to determine if their joint act of price hikes was illegal. There were also the 2011 incidence of Taiwan's top 3 dairy companies having a joint price hike on fresh milk (the result being that the top 3 companies owning a combined 80% of Taiwan's dairy market were fined NT$30 million in total by the government) and the continuous joint price fixing by CPC and FPC for 22 times between 2002 and 2004 (with both companies being fined NT$6.5 million each).

Some of the above cases are about shops having the same owners while some others are about various entities hooking up as market makers. These enterprises as daily services/goods providers can often be seen joining forces to increase their profits, and the general public can only respond by comparing prices or falling into shopping frenzy. But what if these companies on a monopoly strategy have long infiltrated deep into the bottom of the social structure and capitalist services?

Radical Markets in Taiwan:
Extended Reading with Local Perspectives

There have long been conspiracy theories spreading the claim that the American society and capitalist structure are controlled by a small number of families or enterprises. Today, Chapter 3 in *Radical Markets*, finally unveils some of the truths about this rumor. The top six banks that control all the monetary services in the US share many of their top 5 shareholders. These banks that appear different for carrying different brands and strategies and even competing against each other for customers are yet jointly owned by a smattering of holding companies. Such revelation from seeing the chart in the book for the first time hit me like a bolt of lightning akin to the memory of wasting my youthful days in Guanggua Digital Plaza and being conned out of money by the same bosses.

"Since the late 1980s, BlackRock, Fidelity, Vanguard, and State Street have not just grown large in absolute terms. They have also become the largest shareholders of major US corporates.
The top 5 shareholders of the six largest US banks:

JP Morgan Chase / BlackRock, Vanguard, State Street, Fidelity, Wellington

Bank of America / Berkshire Hathaway, BlackRock, Vanguard, State Street, Fidelity

Citigroup / BlackRock, Vanguard, State Street, Fidelity, Capital World Investors

Wells Fargo / Berkshire Hathaway, BlackRock, Vanguard, State Street, Fidelity

U.S. Bank / BlackRock, Vanguard, Fidelity, State Street

PNC Bank / Wellington, BlackRock, Vanguard, Fidelity, Barrow Hanley"

See if you can spot the name of a shareholder that appears only once.

Many readers may recall the term, "antitrust law" which seemed to have protected the world aiming to dismember the octopus (the monopoly) that single-handedly dominates the service market. Indeed, AT&T and Microsoft's IE browser have all been investigated under the antitrust law for monopoly. The antitrust law achieved great success in 1984 in that AT&T was dissembled into an offspring of the parent company, namely, a new AT&T, plus 7 local telephone companies. In the late-1990s Microsoft case, we can see that the antitrust law enforcement came across a setback: in 1998, the US Department of Justice filed a charge against Microsoft for compulsively embedding the IE browser in its operation system which had Microsoft entangled in a lawsuit for antitrust law violation for many years. A settlement was reached in 2001, which was to some degree a triumph on the part of Microsoft.

In recent years, the US Department of Justice has been targeting Facebook, Google, Apple and Amazon with stringent antitrust law enforcement, particularly in 2019. Look at the recent large merging cases, e.g. AT&T merging with Time Warner; Disney merging with Twenty-First Century Fox and Marvel Studios. They are all mega merging projects leaving us exclaiming in awe. In the past, we could hardly imagine that these characters like Woody, Elsa, Darth Vader, Iron Man, X-Men, Avatar, and the Simpson would go under the same company as the Mickey Mouse. The overall value of the intellectual property (IP) of these fictional characters is said to be on a par with the wealth of a nation. The US Department of Justice was at some point concerned about Disney's merging projects, but eventually filed no charges. We can see that the worldwide trend has been consistent with the conclusion proposed by *Radical Markets* authors: the anttrust law, with its good intentions, was once formidably functional but is now no longer a force to reckon with.

"Because of this activism, American antitrust law became a model internationally: it spread first to Britain and then to the European continent and farther around the world. Yet just as American authorities gained the admiration of the world, they stepped off the Red Queen's treadmill. Beginning in the 1970s and accelerating from the 1980s onward, antitrust authorities lost track of the ways in which capital markets reconfigured themselves to maintain monopoly power. In order to understand the reasons why, we must examine the evolution of the corporate form and its governance in the United States during the twentieth century."

Here we are today in 2020; Facebook, WhatsApp and Instagram have close to 3 billion users in total, and all three companies are governed within the same decision-making system. Put simply, Facebook alone has control over the top 3 out of the world's 10 largest social media platforms. In what appears to be a competitive industry, there is essentially little or no competition. These mega enterprises once again act like an octopus gripping the economy and the society, monopolizing any market they want to tap into. This has solidified the power of those possessing contemporary capital and will drastically diminish consumers' right to speak up. The inexplicable disappearance of data as power as discussed in Chapter 5 is also to do with monopoly as illustrated in this chapter.

"Walrus saw private monopolies (along with private land ownership) as both the primary impediment to the operation of free markets and the central cause of inequality, writing in the 1890s, 'Look in American for the sources of the enormous fortunes of multimillionaires... and you will find...the operation of business without competition.'"

Do you want to live in this kind of world? Do you want to end up in a world where everything is owned by the same bosses like all the shops owned by the same people in the Guanghua Digital Plaza? Do you hope to endlessly compare prices just to save more money and to consume more efficiently, when all the prices are jointly fixed under the table by sellers to exploit the rights that you are entitled to as a consumer? In an era of global management and digital communication, we can foresee that the service platforms for industries such as entertainment, finance, science and technology, airlines, automobiles, transportation and accommodation will further expand (e.g. Uber and Airbnb) to such an extent that the all-gripping, all-linking octopus can more easily hide behind all the massive information overload.

One of the authors, Eric A. Posner, is the eldest son in a family of legal scholars (Eric's father being the most-cited legal scholar of all time based on The Journal of Legal Studies). Posner is tremendously intelligent with outstanding employment history. After much deliberation, he proposes in the book the idea of resuscitating antitrust laws and tries to remind us via the term, "Monopsony" that monopoly does not appear only in Wall Streets or indeed on any streets but also in the job markets. Have you ever thought about why wages for the middle class are slow to rise, while the upper-income class often enjoy a quantum leap in their salaries? The above is actually an antitrust act. Those enterprises that pay for our labor can effortlessly achieve their goal of saving labor cost simply by passing joint austerity policies. And why would they not, if they are not being confined by compulsory law enforcement.

Radical Markets: Uprooting Capitalism and Democracy for a Just Society
By E. Glen Weyl, Eric A. Posner.,
Chinese Translation by Zhou YiFang
Chinese Edition published by Gusa Publishing on 2020/04/29
Extended reading for Chapter 4 of the book

5. EPILOGUE: EXTENDED READING WITH LOCAL PERSPECTIVES

After reading this book, I was unsettled beyond words.

Numerous words flashed through my head. If there is to be a new ism, what word can best describe it?

Infosocialism?
Telecommumarketism?
Socialmarketism with free-flowing information?
Or Liquid Democracy

In Chapter 1, I saw resources (assets) liquidity.
In Chapter 2, I saw democracy (votes) liquidity.
In Chapter 3, I saw talent liquidity.
In Chapter 4, I saw power liquidity.
In Chapter 5, I saw data liquidity.

The more liquidity there is, the more freedom and equality there will be in the world which will tip towards the ideal prospect as proposed in yesteryears by those holding free market-oriented views: the market will automatically drive towards an overall optimization and equal allocation.

If the market is to become what the authors shockingly predict it to be in the epilogue: the market will perform as one computer, and liquidity as algorithm, then we can imagine that today's market will morph into a gigantic motherboard or IC chip of a massive volume and size, owing to the world getting flatter with hyperconnectivity. But the liquidity volume of market information along with its connectors (persons and entities that hold such knowledge or information) is somehow diminishing, compared with the past. The market information volume and complexity is on an exponential growth, while people's information handling and comprehension ability is on a linear growth. This will result in a peculiar phenomenon where the minority complies with the majority, and the majority is controlled by an even smaller minority.

What can be done?

Build a computer that belongs to everyone. The computer here holds two connotations: one being an actual computer, through which humans will for the first time have control of a documenting system and fair smart collective decision-making strategies and it will be sufficiently trustworthy, transparent despite its currently insufficient efficiency. The other connotation is about building a market that genuinely belongs to everyone, instead of having a small number of people singularly controlling or even monopolizing the flow of information and decision making

In the end of the book, the authors let their imagination fly and plan a nearly all-knowing almighty (and radical too) computer. Given their economic calculation, they hold the view that with such a huge computer, there will be justice hereafter in the human society, or phrased differently, equality in the world. Every manufacturer anywhere in the world will be informed at all times the demand status of those in need on the other end of the world and can better adapt to the weather conditions in source regions somewhere else so as to timely adjust manufacturing quantity, supply and prices. Viewed from another angle, this computer is no different from an AI mastermind in sci-fi fiction.

Is such a future good or bad? Do you want to trust a super AI computer that anyone can log on to debug at any time? Or do you choose to believe in the country leaders elected by each country via Populism-influenced voting in our time? I have no answers. But I do believe that as the world carries on evolving as it is today, we will surely develop the aforementioned computer, which however may not belong to us the people, and the so-called market will eventually devolve into a mirage for an indefinite period of time.

6. FOLLOW COLUMBUS IN THE MOMENTS OF HISTORY - A FOREWORD FOR TIM O'REILLY'S NEW BOOK *WTF? WHAT'S THE FUTURE AND WHY IT'S UP TO US*

Originally published in 《WTF? WHAT'S THE FUTURE AND WHY IT'S UP TO US》 by Tim O'Reilly - Taiwan Mandarin version

A Legend of Time

Tim O'Reilly is the founder of O'Reilly Media, unique among tech book publishers for its iconic animal covers. Tim is a strong advocate of free software and the open source software movement. He invoked the term, "Web 2.0," and has been propagating the forward-looking notions such as "Gov 2.0." Tim is my idol. I am very grateful to CommonWealth Magazine for inviting me to write a foreword for the Chinese translation of Tim's new book *WTF? What's the Future and Why It's Up* together with **Kai-Fu Lee (Chairman of Sinovation Ventures)** and **Jung-Ting Yeh (Chairman of FamilyMart).** At the end of the article, a discount code is specially offered to my readers for use on Eslite online and Pok'elai. Don't miss it!

Meeting Rooms Are Think Tanks

I often share with students how to train themselves to predict the future and guide them through the way how great creators think. The method is very simple, and I call it "thinking in a meeting room." Imagine a conference room full of people who made history. In this conference room, they brainstorm for hours and discuss how to create something that has been created today.

Employees of Pixar meet to have an internal discussion on Toy Story. (*pixar.wikia.com*)

Through this method, you act like a retrospective detective or disassembly engineer who goes back to and guess at the historical result. Because you already know the result, it is possible for you to compare the result and infer the possible process and important decisions made in that meeting. Although it is a little bit difficult, it is definitely easier for you than for people who are actually in that meeting room because they don't know the result!

I call the meeting room having popped up in my mind many times that gave birth to iPod, iPhone, the first 3D animation, Facebook, Amazon, and the movie, The Fifth Element "that meeting room." The more stories, history, and details in "that meeting room," the clearer we look into the future. We can even create and enter such a meeting room to make history.

The book intrigues me right from the first chapter. Every paragraph is like "that meeting room" constantly reappearing and reorganizing.

Origin

When the term, "Open Source," did not come into existence, the author explained how he had discussions with a group of pioneers in the technology industry and created such a huge historical movement that has shown no signs of slowing down to this day.

The author also explained how he and his friends gave birth to Global

Network Navigator (GNN), the first content-based website acquired by a company in the Internet era, in the early 1990s when Yahoo was still in its infancy or precisely dated a private e-mail to Jeff Bezos, the founder of Amazon at **10:03:59 AM on Wednesday, January 5, 2000.** This e-mail eventually gave birth to BountyQuest, further inspiring Kickstarter and other crowdfunding platforms around the world.

Steve Jobs and the Mac team have a lunch meeting. (*cultofmac.com*)

When I read this book, the secret moments that have such a huge impact on our modern lives are unveiled. **That/those meeting room(s)** must exist. We just have to delve deep in order to enter. All the great things, as they are, have a beginning. This book is full of awe-inspiring moments when you find out that something just looked differently at that time, and the author played his part in those moments!

Moments of History

After I finish reading the book of nearly 500 pages, it seems that the author guides me to a place between the present and the future. A miraculous office building nestles in the time corridor, where celebrities from various countries appear to work from "that meeting room" to a full floor office space.

When walking down the time corridor, sometimes you will encounter celebrities such as a ministerial official from Germany in the G20 meeting criticizing Uber, legendary investor George Soros talking to the author about the "post-truth era," Washington Post columnist John Farrell debating on fake news, the CEO of General Electric and the White House administration expressing their views on the next economy to the author at the NEXT global economy conference. In a certain chapter, we can find an example from Taiwan, vTaiwan, led by former minister of state Yu-Ling Tsai, Audrey Tang, and g0v (see g0v.news), which is the best case for the author to explain how **technology is changing politics**.

vTaiwan is a virtual policy exchange platform and a model for digital governance in Taiwan. (*vtaiwan.tw*)

In the last chapter, the author talks about his views on the future. In general, he is optimistic about the future, but he also expresses his concern about the current situation. In addition to concerns and discussions about **"super currencies,"** he also explains and encourages **"universal basic income"** (see the article by Jia-An Chu, a writer of philosophical thinking); in addition, he also points out some unique ideas. For example, he proposes that "future currencies should be divided into two types: **"machine currency"** and **"human currency."**

Reflection

Before war, soldiers carefully read the map lying on the table. Likewise, the author details the frontier technologies over time, trying to lead us to advance and adventure. This map constantly reminds me of the blockchain industry that I have been engaged in recently. Blockchain technology was proposed a decade ago, but there are still many terms that are undefined (e.g., DApp, and Backup Phrase). Comparing to the trajectory of technological developments described in the book, it is like a recurrence of the author's participation in the creation and development of the Internet. This makes me feel more confident of developing blockchain technology, and many scenes and experiences described in the book become the best cross-temporal comparison and reference for my current work.

Conclusion

What our future will look like depends on how we take on the predictions to shape this century. In the past, we might be hesitant and kept in the dark. After reading this book, we will all become Columbus in the moments of history, holding a map full of history, experience, stories and courage that guides us to **a more human-centered future** depicted by the author.

Originally published in 《WTF? WHAT'S THE FUTURE AND WHY IT'S UP TO US》 by Tim O'Reilly, Sky Magazine, translated by Ting-Min Huang, 2018/12/26 Published.

Radical Markets in Taiwan:
Extended Reading with Local Perspectives

7. BBFT PODCAST #EP2 "A MEETING IN THE FOREST TO DECIDE THE FUTURE OF HUMAN FINANCE!" FT. DR. TOM LAM

Originally on SoundOn & YouTube
https://youtu.be/tLQaUdBAx7s

It may be a tough day for you listeners, and today's show may be a bit hard. Although Dr. Bao has always wanted my show to have you laugh more than you gawk. Or more time to understand than not, but today is a challenge for you. I hope you can try to listen to a very sexy Cantonese accented Dr. Lam talking to you about the thinking and reasoning of an economist.

Technology, innovation, entertainment, all kinds of novelty, interesting are in Baobo friends said, you listen to Baobo friends said?

Dr. Bao: Hello everyone listening to friends, I am Dr. Bao. Today we are here again with this program from our friends at Dr. Bao's show. No matter which platform you are listening to this program, you are welcome to continue to listen to it, do not change the channel oh. Today, Dr. Bao is a little bit nervous again because today's guest is a very rare and special guest. He took a flight of 11 to 13 hours, depending on which route he took or which coast he departed from. He came from the United States to Taiwan to attend the premiere of the Saint of Thieves movie. The co-founder and chief economist of SELF, Dr. Tom Lam, is here. Let's give him a round of applause, and let's say hello to the audience.

Lam: Hello, everyone.

Dr. Bao: Do not listen to him as if he is very modest, there is no such very strong voice. His mind. In fact, he's very powerful. When Dr. Bao first learned about Bitcoin in 2013, the only person he could ask was Tom. Because he has a PhD in economics from the University of Chicago. You know, there are so-called schools of thought in economics. One of the big schools, I don't know if it's the biggest school, is the Chicago School. What is the first big one? The second largest? Like this Kunlun School, how is this illustrated?

Lam: You ask someone who graduated from the University of Chicago, of course, they'll tell you they're the biggest. But if you ask a Harvard graduate, they will also say they are the biggest.

Dr. Bao: So that's why Harvard are the second faction, right? Are they called Harvard?

Lam: No no, names are given by others. You know you call yourself by your own name, it's usually a fake. Usually it's given by someone else.

Dr. Bao: At this point, you should have noticed that Dr. Lam's accent has some very sexy Cantonese accent.

Lam: I just don't speak Mandarin.

Dr. Bao: No, I think it's very standard, no communication problems at all. Dr. Lam's current job. In addition to this identity, he also has a daily job with Superman, which is an assistant professor in the Department of Economics at Clemson University. Like Dr. Bao, the most in the academic world. The most youthful, the most youthful years, with all the most eager to cooperate with such a class. You may not find it funny.

Lam: People in academia will laugh.

Dr. Bao: Everyone in academia laughs. I'm not sure if it's a good idea, but it's a good idea to have a chat. Tom is actually relatively low-profile on Facebook. So we are in the private message chat. So recently said that there is really no chat, so we are not necessarily specific to talk about what topics. I think economists are a rare breed in this world, right? Isn't it? Overall. That is, which is more common, computer scientists or economists?

Lam: I don't know, yeah. But I think, because you are more computer science friends, you feel that there are a lot of people in computer science. I think I also think economists are full of, too much.

Dr. Bao: So for me, I think it's great that someone in my world got to know a real economist. At that time, I couldn't even understand Bitcoin, so I was the first one to send a message asking him what was going on.

How come there is a bitcoin that can exist without a government or a country? Is he really not a scam? And so on and so forth. Let's not talk about that, because this blockchain that the cows will also be in doesn't necessarily start officially today, anyway, it is or it doesn't necessarily count as an official episode. Because it is rare to fly in, we have to ask the economists what they think about this world anyway. I actually did a little bit of research myself, and I just listed some of the topics that I wanted to talk about. I heard that you just read the first question you have an opinion.

Lam: Yes, I have super opinions.

Dr. Bao: What is the first question? As an economist, how do you manage wealth. I didn't say that you're going to advise us on how to manage wealth. (05:00) We just want to know how you manage wealth that way. Or what do you think is the way to manage wealth?

Lam: The way economists manage wealth is to find professionals to manage it. Do not look for economists to manage.

Dr. Bao: So the word economist and wealth are two different conceptions.

Lam: But it is also possible that some economists are also very good at managing wealth, which is possible.

Dr. Bao: What about the ratio? How do we test whether an economist is an economist who can manage wealth or one who can't?

Lam: Let me tell you, many economists are very poor. So it doesn't mean that finding an economist to manage your wealth is particularly good.

Dr. Bao: That there is no test question to test his sentence is to know that is.

Lam: Do you know how he is, will he manage the wealth?

Dr. Bao: Is there such a thing? No?

Lam: No, I guess. I was thinking, because you asked me how I manage my wealth, and I don't know. I should say that everyone has their own way of managing their wealth. I don't know if I'm a professional or a good one.

Dr. Bao: So in short, many economists are very poor.

Lam: I tell you one thing, you talk to those who study economics, you will find that they will have a trouble, that trouble is what? For example, you go to study a doctorate, or you go to a famous school to study. Then, your relatives and friends will come to ask you what questions, they will ask you which stock will rise tomorrow? Which stock will go up tomorrow? Then people who study economics will be super, super confused. I think, I do not know.

Dr. Bao: Everyone will not be able to resist it.

Lam: People won't be able to resist.

Dr. Bao: I also admit that I couldn't help but ask Tom a few times about that what coin.

Lam: What currency will rise, you ask those who speculate in currency.

Dr. Bao: I am more polite, I will say how you look.

Lam: Yes, yes, yes, you will say what we say about academic research.

Dr. Bao: Okay, so in short, economists and wealth experts are two different kinds of people, so don't get me wrong. So if you're curious about how an economist manages wealth, you might want to ask him which bank he has an account at first. Is there a recent track of this buying and selling ** (06:59). But usually you'll probably be disappointed. According to this Tom from the economics school, maybe we ask the Harvard school and they choke you. Say no, we all drive Tesla.

Lam: Yes, yes, yes, you ask the other may be very powerful. He can say that I do not understand, it is possible.

Dr. Bao: If you are an economist, or you are an economist friend, and you do not agree with this matter. You are welcome to leave a comment at the bottom. Anyway, the next question we actually want to talk about, this one is also voted on. In fact, we only have two votes. It's me and my wife. These two questions are what my wife wants to hear. She said she wanted to know how economists judge the future economic trend. I think this future is a bit too long. Let's give him a 30-year time frame. Do you have a chance in 30 years?

Lam: In fact, the future of that period is not so important. You say ten years or thirty years, it doesn't matter. But the question is which part? This is the most important.

Dr. Bao: So let's think about it. I know, I know. I've been reading a lot of reports lately, and just friends have been telling me, that the world is going to enter an era of negative interest rates. It seems that Japan has started first. Is your money in the bank inside, will become less and less. Because he does not want you to exist in the bank. You deposit a thousand dollars, tomorrow becomes 990, you put again. I'm just exaggerating. Just say your money will be saved more and more to save less and less. Forcing you to do something else. I heard that after that, many other countries will successively put in this kind of negative interest rate behavior. Then, for the first time in human history, the world will enter an era of negative interest rates that may be a joint effort. Is this true? What do you think?

Lam: That would be the case. Why is there such a thing as interest rates? The interest rate thing is basically saying, I want to get a money today, I don't want to get a money a year later.

Dr. Bao: That is to say, it is all fishing. If there is no incentive to sell the money earlier, why should I go fishing earlier? Right?

Lam: For example, there is a person, we hope, on the government hopes that people go to invest a little bit more. Half of them want to invest, but they don't have the money, they just have a good project. he'll borrow money back, right. Borrow money to come back.

Dr. Bao: building houses, building high-speed rail, zoos.

Lam: Make a movie or something. And then, if they can do it, that's the best thing. But what if there is no money? He gives a little meaning, he can borrow the money back.

Dr. Bao: That's me, if you lend me a hundred, the next time you change you a hundred, you will have no incentive. So we have invented something called interest.

Lam: Yes, yes, yes.

Dr. Bao: Interest or interest rate.

Lam: It makes to lend to others, can get some return. But the problem is that this thing is, is to make that investment or business people is the cost to come. So the best.

Dr. Bao: You said the people who make movies. So if the negative interest rate I borrow 10 million from you, and then I pay you back 9 million. I said negative interest rate. (10:00)

Lam: Do not say negative, zero, such as saying zero is also very happy, on you borrow money.

Dr. Bao: Who is happy? People who make movies are happy.

Lam: After I borrowed the money, I changed the same money. For that investment environment, you have a low interest rate is very good.

Dr. Bao: But he is now negative.

Lam: Why would it be negative. The government has been doing a lot of things to suppress this interest rate, this thing. Why often people will say that there are some countries will print money. Printing money means, you know, a lot of money on the market.

Dr. Bao: We will not be lied to? The government has lied to us and said it's okay. We will negative interest rates negative a little, the more negative the more suddenly the world into a super strange state is.

Lam: What is the super strange state? It's that you have to keep interest rates very, very low, and there is a price to pay.

Dr. Bao: What's the price?

Lam: One of the methods is that you print to everyone the whole market above a lot of money, money can be printed out, will cause some problems. But he will tell you no problem before causing problems.

Dr. Bao: I understand it la. I think this academic community, after all, they can not speak too, you know, they need to be responsible for their academic reputation. So he can not way in front of a large audience, especially tens of thousands of Bao Bo friends said the audience, to everyone speak out too alarming such prophecy. Otherwise, it will become what Mr. Wang said next June will be what earthquake, the end of time so. If it does not happen, people will leave a message at the bottom to scold Tom. this I understand it, I hope you can hear, listen to the meaning of the words of Tom Bo so. We should not ask him again. Well, these days there is in the release of this Nobel Prize winner. I see to a news is to say that this year's physics is also the winner of the chemistry prize, before the announcement. That Nobel Prize chairman or a person who released the prize, it also read a paragraph of that Big Bang Theory, the Chinese translation called what the "Big Bang" or what the "geeks can do" of that title. This is a very hot universe that paragraph I will not sing. In Taiwan, someone wrote an article. The Minister of Science and Technology, Minister Chen Liangji, also reposted that in fact, "The Big Bang Theory" is really a great program because he used a science program to let a scientist play Amy, the geeky girlfriend of a geeky scientist inside Sheldon, a real neuroscience doctor. And then let him play a neuroscience doctor. Then, it is a scientific joke funny this kind of daily program, in fact, encouraged a lot of young people small children, want to become a scientist. Because it looks very funny, very funny, very funny. I was thinking of saying that you feel like an economist. The first is that you think economists often do not often on the show?

Lam: Not often enough. I think we really should have more.

Dr. Bao: Is there any kind of podcast for economists in the US?

Lam: Yes, but some of him are very academic.

Dr. Bao: It's not funny, there's no kind of plot packaging or anything.

Lam: science popularization of such things. You say "The Big Bang Theory" he is not saying to you to talk about a lesson. In fact, it is still a drama to come, or can have a lot of very funny plot. He has some things that can be popularized in the middle.

Dr. Bao: That time we Xu Jia Kai director, "The Saint of Thieves" Golden Horse Award new director Xu Jia Kai. After we come to engage in a web drama is the economics of the great convergence.

Lam: No. Another thing is that in fact, many of my colleagues or friends with some other colleagues are economists. They all said, they said those things are very complicated, what, it is very difficult to use an easy way. I said, "Actually, this thing is difficult, and science is difficult. Not that we can easily do science popularization by just saying it. However, it should be possible to do. If physics can do it, why can't economics do it?

Dr. Bao: The most famous one in the US is called Neil Tyson, isn't it?

Lam: Neil DeGrasse Tyson vs.

Dr. Bao: He is a everyone to see the star effect that movie, everyone began to have interest in space black hole. The United States on a bunch of programs is to find this Tyson to go on the program, to talk about this science, also triggered a lot of people's interest. That Big Bang Theory also has this physicist, then there are space scientists, neuroscientists is also a lot of scientist. this economist you this circle, the

Lam: There we have The Big Short.

Dr. Bao: big short sale. Just a two-hour look at the end of it, right.

Lam: Just trying. There should be more in the future.

Dr. Bao: So I think we can get it. Then if director Xu Jia Kai wants to do it, he will ask you to play the neuroeconomist.

Lam: He could have come to me, too.

Dr. Bao: It's a little nuts. In fact, this brings me to my next question, which may not know whether it will be a bit serious. That is, in fact, I have recently been feeling (15:00) in the world of blockchain, a lot of the original economics, very basic, fundamental, fundamental principle is that no one in this world has ever talked to us. No one has ever told us where the money comes from? No one has ever told us that printing out a banknote needs to go through what procedures, and what price this society has to pay? And no one has ever told us how we people changed from the age of shells to the age of gold, how the age of gold became the age of gold coins, how the age of gold coins became the age of banknotes, and how the age of banknotes. We just find that we don't know anything about it. When we see the block chain, we say, "What is money? Oh, this is the transparent cup, this is the verifiable book, everyone is dumbfounded, what is this?

Lam: The reason is because you don't have, originally you don't know what those one by one bills are for. You think you know.

Dr. Bao: You always had a cool story about you opening that Hong Kong dollar? Hong Kong dollar. The Hong Kong dollar bill with what it says on it

Lam: Hong Kong dollar that is, you open an old Hong Kong dollar words, he will write a diploma, for example, say twenty Hong Kong dollars. He will write the voucher quality to pay twenty dollars. He said you are a ticket, take this ticket I will pay twenty dollars to you. People are super strange, you are not already twenty dollars?

Dr. Bao: I'll take it to the bank and exchange it for another one.

Lam: This ticket, you promise to exchange it for twenty dollars to me. This is not very strange? This is the original concept of banknotes. He can pay a thousand taels of gold with the ticket. You can exchange it. But his advantage is that you use that paper after. You can't take the gold with you because it weighs tens of taels. Then you can bring the banknotes to and fro, very easy, so the transaction will be very convenient. But in fact, he actually or behind him is able to exchange those money.

Dr. Bao: He himself. Think about it, people. You a hundred thousand dollars bill, does not mean that this piece of paper is worth a thousand dollars. He is someone in promise to say, guarantee that you can always take this thousand dollars to buy, I am able to give you a thousand value of something. Maybe it's a coin or maybe it's gold.

Lam: You use an idea of words, he is not necessarily gold. You can also be a ticket to pay for a meal.

Dr. Bao: A thousand dollars for the meal.

Lam: Yes, a thousand dollars of rice.

Dr. Bao: But the point is that in the world of blockchain, we'll find that. When the person who gave you this ultimate guarantee reneges, or cheats. That is, he printed a lot of tickets himself. Anyway, what squeeze, the country decided to issue new money. A few days ago, India was doing this trick. The old banknotes cannot be exchanged, and everyone has to move out from under the bed.

Lam: In fact, many countries have tried this.

Dr. Bao: A certain place very close to us, too. There are many things that we think are natural. Human society has this thing.

Lam: The reason is because of this, at the beginning of the development of things, at the beginning are the kind of ticket for gold or for a meal, we all think it is very simple. Usually a simple thing at the beginning, people will make it complicated. The more complicated it becomes, the more complicated it becomes. People forget how it came to be in the first place. When it becomes complicated, everyone will have, what is called cheating.

Dr. Bao: Because there are some systems you can't understand, you can't see.

Lam: People started to get confused. Then the smart people, the relatively smart people started to profit from it.

Dr. Bao: inside the opaque cup to change the magic. I have been talking about the cup, the reason is that you can go to Google "Saint thief" of the introductory trailer. That is, we are talking about the current flow of money, like changing that cup magic. You say, where is the money? It's not there. You now go to the bank inside to find your deposit of a thousand dollars. The bank can't take out the $1,000 that you originally deposited.

Lam: No way.

Dr. Bao: And he took out a thousand, or maybe yesterday Tom saved into a thousand. Not I deposited into the thousand. But the bank has asked Tom said, Tom sorry, I took your money to Bao Bo hello.

Lam: What if I want to take the money? He will take another person's money to compensate me.

Dr. Bao: Right. So, because why there is a bank reserve ratio of 20 percent, and each country also each era is different. So this is, I never knew. I told you, I never knew. I'm a doctor, I don't even know. I'm sorry. I just don't know. Not to mention, that is, if you understand the matter, you will find that money has some properties, definitions, and a process of its birth that we do not know.

Lam: The most important thing is that all these things are okay at the beginning. He will tell you a very reasonable. For example, the bank to take your money. At the beginning, in fact, did not say unreasonable. But when it becomes more and more complex, it will be those unreasonable things mixed in it.

Dr. Bao: Just change the cup trick.

Lam: Right.

Dr. Bao: Anyway, his eighty cups are over there like this.

Lam: You don't know the reason either.

Dr. Bao: You don't know where my money goes, do you? Wait until I say I want my money. He said, then you turn your back, wait 30 minutes and I will find out to you.

Lam: (20:00) All things are not in trouble until they are not in trouble. This is.

Dr. Bao: Do not say more, do not say more. I think it includes the country, what is the country? I actually also heard Tom tell me a cool concept. She said that in economics, the concept of the state, in fact, is called economic is an economic body.

Lam: Yes, an economy.

Dr. Bao: The emergence of national borders may well be related to the limits of the circulation of a currency, or the range of movement of the military before, and so on, which is very complicated. But, I was actually thinking at the time I asked him to say that I wanted to promote the idea that Taiwan could start teaching economics in kindergarten afterwards. I'm talking nonsense, not kindergarten, maybe elementary school. Our school earth science, Mandarin, English, American labor, physical education, we use money every day. Elementary school students now have to go buy ten dollars of bread. And then everyone is arguing about the smart vending machine. The economy is right or wrong.

Lam: I do not know in Taiwan, in Taiwan you learn economics in high school?

Dr. Bao: No ah I did not even study in college I tell you, I am an economic idiot.

Lam: In Hong Kong, we teach economics in the third year of junior high in most schools.

Dr. Bao: junior high school third grade ah? Junior high school seems to be Taiwan's may be national high school, right? You see we have so many retards in Taiwan. I speak for myself, I speak for myself. I really feel like talking about what to do? Every day, my right to live, my job salary, why is it low? Why are these structures created? I don't know anything. We are just saying that we are angry and we are going to the street, right?

Lam: I probably agree with studying economics earlier, but it depends on how he teaches it. It depends on how his curriculum is set. But in principle I do agree.

Dr. Bao: It's popular in the US, the earlier you teach something new. Like that they now computer science writing program, they have that kind of quantum physics for kids. is quantum physics for young children.

Lam: I think this kind of also has python for.

Dr. Bao: What program language right or wrong, how to do the site three years old.

Lam: Usually when people ask, I have to see how he does. Some are not doing very well.

Dr. Bao: Kids are bad.

Lam: Because they don't really want to teach, they just want to be a *** (22:23). Sometimes you are instructed to say that there is a title that says you learn when you are five years old, before there is a five-year-old, MBA? Just mess around.

Dr. Bao: The Chinese love this set. In my school, I said I started teaching when I was eight years old. Another one says, "I started teaching when I was seven.

Lam: I didn't say Chinese, which actually Americans have. People have that a lot of the time. If you really do this kind of course well, there are some concepts that are really helpful to learn a little bit earlier.

Dr. Bao: Yeah, so.

Lam: For example, everything has a price, this kind of thing is also a concept of economics, I think it is good for children to learn.

Dr. Bao: All in all, I feel like I'm a conspiracy theorist. Sometimes I feel like people are deliberately not teaching us. So we are right or wrong. I don't know what's going on, why my money is missing from the bank. I actually want to promote this, but I don't know where to start.

Lam: Just now not say science popularization. I think a lot of them are, you do not teach in school, you will be taught in science. Popular science teaching is actually a good thing, we do not have to take the test. Like to learn can learn more. You use a fun way to learn, but also can make up for some of the school curriculum deficiencies.

Dr. Bao: It's true that we made the film "The Saint's Bandit" and came back to put it in. The movie "The Saint and the Thief" is really an attempt to do something. I think the film makers are also a little bit nervous that people will think that the blockchain movie is scary. I think it's even scarier than "Back to School", so I don't dare to go in.

Lam: Obviously "Back to School" should have been scarier.

Dr. Bao: Yes, yes. No, but "Back to School" was very scary and then people went to see it, and then felt that the block chain was very scary. Everyone went to see it and then found it so scary. Recommend everyone, you know that the "saint thief" super scary.

Lam: Not bad this way. At the beginning of this film to me, I am an economist, I am not making a movie. I also think that you are shooting a film, and then found that your film is very serious. So just want to find an economist to say that it is not just a movie, but also a little something.

Dr. Bao: I've said that the blockchain is the world's Ju-light garden. You may not listen to understand. The boys went to the military to see some very, the plot is relatively no ups and downs.

Lam: I understand this.

Dr. Bao: Compare some of the training and education films. But we are very wonderful, we are very wonderful. Wonderful on the explosion of wonderful.

Lam: It's important to be wonderful.

Dr. Bao: Talk about a gossip, that is, I heard that the movie "The Saint of Thieves" has some images, that is, some economic formulas will appear on the wall. We do not know if you have seen a movie called "social networking" is talking about the process of the creation of Facebook. There is a classic picture inside, that is David Fincher, directed by Leonardo da Vinci, I remember (25:00). In that transparent window, write on some formulas, look more geek. then is wow write some very powerful mathematical formulas, talk about how strong we this Facebook algorithm, is fierce over the whole Harvard University. We also have a similar plot, that is, a very powerful start-up company, invented some new things. There are some formulas on the whiteboard, if you have a look at some companies. Of course, if you have watched that Big Bang Theory, geeks big bang, life can not. Inside there are often such eggs in the back of the whiteboard, or on the wall, above the blackboard. Heard that the saints, also have. Each of his formula came from Tom's hand.

Lam: I wrote it all.

Dr. Bao: You actually don't know which lenses yet.

Lam: I tell you, write the formula on top of the glass this thing, and then find the movie also has. It's Beautiful mind.

Dr. Bao: Taiwan flip beautiful realm. A mathematician.

Lam: They have too.

Dr. Bao: There must be a very focused on holding a strange pen to write on the glass, why because luck mirror is better.

Lam: Yeah. You're hard to shoot on the blackboard.

Dr. Bao: You can capture both the words and the people at the same time. Did you write a lot of formulas?

Lam: Just write a lot.

Dr. Bao: really.

Lam: Because I also do not know where those will be, on which one plot appears, I do not know. But I probably know is used to do.

Dr. Bao: So you wrote a bunch of formulas and then drew a bunch of diagrams for him.

Lam: Yes, yes, yes.

Dr. Bao: Then you also did not do note, so it is possible that the protagonist in the discussion of interest rates on deposits, the results of the whiteboard above the talk is that.

Lam: I did explain it to the director.

Dr. Bao: So no economist will look at it and think it's not right.

Lam: I don't think so, I don't think so.

Dr. Bao: Anyway, I heard that there is this egg. Is every formula verifiable by the Chicago School?

Lam: This has nothing to do with the Chicago school. This is, because you have a lot of actually are from some thesis also have my own written. A lot of them are written by myself. Because in that thesis, I will certainly check, know that he is right before writing up. So it should be no problem.

Dr. Bao: I heard that you have to leave a hand. You have hidden one is, you have to really be an economist, you look very carefully, you may find a bug, really have this bug?

Lam: No, there are some places that are.

Dr. Bao: It's obviously two, you write it down as three for him.

Lam: No, no, no, it's not like that.

Dr. Bao: It's obviously X and you write it as Y.

Lam: You have said inside on a new start-up company in writing a whiteboard. If you're real, it's impossible to write it all right. So it's very real, even the part about writing wrong is very real.

Dr. Bao: So in the end, is there an egg or not? If the listener has an economist.

Lam: When I was writing, I did write. I don't know if there was in the final cut out version.

Dr. Bao: Maybe the director of that team crew copy are copied wrong.

Lam: I don't know if he has cut it.

Dr. Bao: So you are welcome to come and find fault, that is to say.

Lam: When I look, I also come to find fault.

Dr. Bao: It was so hard to watch. I hope you look forward to it. We are involved in the well, we will fatten up a little. In this program above. The friend of Bao Bo said, five words have two words is Bao Bo, speak with Bao Bo related is also OK. Well, but we think it is still to talk about my own very serious want to talk about the problem. It is this Brighton Woods Conference. It seems a translation method called the Brighton Woods Conference. I must say, I'm ignorant. It was in 2014, when I was researching blockchain, that I realized that there were a handful of people, like seven people, six people, eight people. One year in the forest, a meeting was held in the forest.

Lam: A place.

Dr. Bao: It's not really a forest. Just like we talk about the fence is not really a fence with wood.

Lam: Don't imagine him in a tree, not that kind.

Dr. Bao: I am feeling that they are hiding in the forest inside, secretly opened a meeting. This meeting inside decided some very mysterious things, do you want to say, this what bird meeting?

Lin： Inside there are a lot of detail, but then I know what you want to say about that is what. I want to say is how to say ah, a policy. is in the end we issued the banknotes, in the end behind to not be gold guarantee?

Dr. Bao: I wonder if you have heard of three very strange words called gold standard? The gold of gold, the original, the position of the position. In fact, it is about the former dollar. In the past, the back of the banknotes of this country must have.

Lam: Commodity currency.

Dr. Bao: Yes. To have a commodity, or a something.

Lam: Something of value.

Dr. Bao: with a note to do matching. Just respond to this before Tom speaks of this Hong Kong dollar. With a ticket to pay for something worth twenty dollars.

Lam: The dollar is also, why is it called the dollar? Because it can be exchanged for gold.

Dr. Bao: The U.S. dollar can actually be exchanged for gold back then.

Lam: You can really exchange gold.

Dr. Bao: take this very beautiful U.S. currency, to exchange for gold, so we call the U.S. dollar. So a group of people hid inside the forest, after a meeting decided (30:00), decided that all the dollars can not be exchanged for gold. That's it. So you can only exchange dollars for dollars. They hid over there and had a meeting. And then they said, "We found out that there's not enough gold. I know that some of them went to the place where they used to keep the gold under the U.S. Federal Reserve in New York and found that they had taken it and lent it to people. And then the rest of it, why the hell is there so little left? Fuck, it's also very troublesome to come back. Then, or this is good, we will cut. It's a bit like breaking up. Call, also can not call to the gold, also can not call to the world has the dollar people, that otherwise we will break up hello. Then we broke up. That's how all of us who have dollars around the world, they lose the 100 percent quid pro quo relationship that they can exchange for gold. Later you will find that the dollar is the dollar, gold is gold, he has a price of his own. I remember one of the people who attended the meeting was a man named J.P Morgan, right? Right?

Lam: I forgot about it.

Dr. Bao: There are some financial giants.

Lam: There definitely is.

Dr. Bao: with the senior government units. It seems that there is more than one country.

Lam: Yes, yes, yes. I didn't remember that list, I didn't remember all those people's names. You can look it up. But I just wanted to interrupt you at that place. I just said, he was originally talking about the exchange, but there is an exchange price, on that gold. Later, he wanted to cancel this thing. Before, he also went through a process, he adjusted a price. On the adjustment of some, on a year on the adjustment of a little. Adjust ah adjust ah, we found no way out, and then said I will not adjust on the cancellation of this thing. It is after a process. But at the beginning you ask this thing when, with a ticket for gold is not important.

Dr. Bao: Yes.

Lam: Now can not be replaced when you look at the United States did not have the whole country rotten.

Dr. Bao: So if he really goes crazy one day, he'll say he's sorry we transferred.

Lam: In theory, yes. For example, the United States often says that it can print money. So it is possible. Theoretically, you can print money indiscriminately. But he didn't.

Dr. Bao: We can only trust that he will not go crazy.

Lam: Yes, you could say that.

Dr. Bao: Let's continue to demonize this first. Not la is I check Wikipedia, he is saying in this Bretton Woods Conference, so is not in the forest?

Lam: No.

Dr. Bao: Not the forest. What to do, this whole atmosphere is destroyed. My mind among them.

Lam: Imagine Lord Ring with some secret meeting.

Dr. Bao: I have a very complete forest structure in my head, and now I'm breaking it down in an instant. And then, in fact, there are American representatives. This Harry Dexter White, White is very famous, he seems to be the former Secretary of the Treasury. Then there was also a representative from China at that time, called Kong Xiangxi in 1941. Why this meeting I find it very interesting. First of all, it does not matter whether it is a forest or not. It is the blockchain world, there is a part of the people very opposed to this meeting. Because they think that you can come at will after you cancel the right to exchange the banknotes for food. So what we have in our hands, the banknotes, are very illusory. We must trust the government not to mess around, so that what we have in our hands has value. At one time in the blockchain world, I even once participated in this angry circle. I thought how this chaos in the forest will open up like this. Later I read a book called "Financial Curse", also a friend of the baby, talking about Brother Liu Yicheng, one of the founders of the future bank. He introduced this book. I finished reading it. This book was written by an Englishman, and it was very strange that he said that at that time this conference was spending a lot of energy trying to save the world's economy. He said that they found that there was a lot of unproductive movement of money. That is to say, I put the money to open a company tax-free paradise those countries such as Salvatore la, what island ah, BVI what is some. Then, they suddenly found a lot of worthless money flow. But because of the tax rate or something, creating a lot of meaningless income, interest rates, ah, or the price difference exchange rate difference. The money just keeps flowing and flowing, more and more. I heard that the forest meeting was to stop this thing, and then it failed. So, after the meeting, many people did not comply with the meeting reached some consensus. As a result, we still have a lot of this money flowing around, not to produce.

Lam: I can't guess what their purpose is, but it's actually pretty obvious. He said if, if you have an agreement is to say that you want to be able to redeem gold. In fact, can exchange this thing is not the most important point. It's the fact that you can't print randomly that's the most important point.

Dr. Bao: But he didn't. He just said I wasn't right enough.

Lam: So there is no promise in the time of fulfillment, we can so-called chaotic printing. But you also have to think about why people would be willing to get off this gold standard this thing. (35:00) In addition to the fact that they simply do not have so much gold to exchange. In fact, after a period of time, many economists will find. If the supply of money is completely fixed, it will be a problem. Therefore, the production of gold, the government is not completely fixed. But you can't increase it too much.

Dr. Bao: Suddenly tomorrow becomes ten times, there is no way.

Lam: Yes, yes, yes. But sometimes for some good purpose in this economy we need to adjust this money supply, so gold doesn't have this flexibility you could say. Back to bitcoin, or virtual currency. Everyone will say, to have a selling point. Bitcoin, especially two or three years ago, everyone would say, "What are the benefits of Bitcoin? One of the key benefits is that the supply cannot be increased, it has a fixed amount.

Dr. Bao: Twenty-one million bitcoins.

Lam: It's kind of like the gold of this world, you've got so much of it. So he can't just send. This is, everyone says this is a benefit. As a value preservation thing, it is a benefit. But as a currency, he is not a benefit? If he is really that good, we can now also use gold right.

Dr. Bao: I actually think that the audience might have a hard time. Because Dr. Lam is a scholar, he's not going to tell you the answer. He is, is there really a benefit to this? It's not that there are no benefits. This really has a downside? It's not without its disadvantages either.

Lam: Let me give you an example. You can imagine a very, very small store, he handed out some breakfast coupons, right. You can take it for breakfast. How many breakfast coupons can he give out? Send too many, there is a problem, right. A store he can not exchange. But if you send too small, it will not have the original purpose of the volume. The first thing you need to do is to get the best out of it. The volume is a bit like currency. So how much of that currency should be issued should be different in different situations. Even if you didn't study economics, you would know that a small breakfast restaurant shouldn't give out too much. If you are a multinational group, you can issue a lot more. The economy is the same, you can't just have a completely fixed amount and say it's the best.

Dr. Bao: I don't know about the audience. In fact, when we were talking about the breakfast restaurant, I was already inside the breakfast restaurant in my brain. I wasn't on that roll. But I understand, in fact, Tom has been very enthusiastic to spend a lot of time with me to introduce this some I can not think of things. He will give many examples. But also hope that the audience. I am open source today, temporarily open source this open source. this Tom, let everyone to listen to see his views on some things. So we just listened to some of the meetings inside the forest, and then or on top of the Hong Kong dollar, on the old Hong Kong dollar, on the banknotes some mysterious language. It was very interesting. Finally, the last thing I want to ask is. Because we have a very important theme in this series, that is, the blockchain called cattle will also be. I would like to ask you, do you think blockchain is really worthy of cows? That is, is there really a way for cows to use blockchain? Do cows really want to know? I mean, you as a blockchain, you can't say player, you can say intellectual. The intellectuals of the blockchain world, the experts in economics. Do you think that people really want to understand blockchain? Do we really understand it? Is it worth spending our brain power to understand it?

Lam: All these things have a technical level and an application level. Application level, I think we all need to understand. Because this is indeed, at least I believe is the future on this world a very important component. But the technical level, for example, the Internet is not a lot of people know what the bottom of the Internet is doing.

Dr. Bao: TCP/IP ah, UDP ah.

Back then, for example, when people said I want to learn the Internet in the nineties, they said I want to learn the Internet. You are now using now twenty years later people to talk to them, you will say with him of course, to learn ah. But in fact, we are talking about the application level you want to learn.

Dr. Bao: When we first started receiving E-mail in the 1990s, we had to set a bunch of messy numbers, what with SMTP. 445 port, then pop3, then 433 something.

Lam: Exposed age.

Dr. Bao: just to receive a letter, an afternoon without this feeling.

Lam: But I believe that if this blockchain will change the world in the future, it will be the same as the Internet. That is, there are a bunch of smart people who will make it easier and easier to use. People can not understand so many technical aspects of things can be used, this should be will.

Dr. Bao: Great. We believe that Tombo, Dr. Lam will be one of the smart people in this bunch. Because he at least know Brighton Woods meeting is not in the forest inside OK. I hope so, everyone.

Lam: Put it this way.

Dr. Bao: We laughed so much ourselves, the audience is also very annoying. Well, at least he knows that the Brighton Woods Conference is not in the woods, okay? We are very excited about this program today, and very happy and grateful to have this Dr. Lam Zhongsheng Lin who flew in from the United States. For those who want to know more about him, feel free to Google his information on the Internet, it is very mysterious. For the end is not Brighton Woods in what kind of place, but also welcome you to spend a little more leisure time, more understanding of the Dr. Bao friends said. If you know enough about it, maybe one day we'll invite you to come up and be our friend and talk to Dr. Bao.

Thank you for listening to today's blog, I'm Dr. Ruth Gerber. If you like my show, remember to subscribe to it on SoundOn. If you heard us on the platform, feel free to give us a five-star rating and leave a comment or click on the little bell to follow the subscription. We are very happy today, and we thank Dr. Lam for coming to our program. See you on the air next time, bye bye.

* Note: Some of the text in this chapter is marked as time code due to dictation problems, so please go to the original release channel for further information.

Radical Markets in Taiwan:
Extended Reading with Local Perspectives

8. BBFT PODCAST #EP5 "THE UNIVERSITY OF CHICAGO GENIUS PROPOSED FIVE MOVES TO RADICALIZE THE MARKET TO SOLVE THE CAPITAL PROBLEM!"

Originally on SoundOn & YouTube
https://youtu.be/_MapxkQN6MI

A little like, I do not know if you know, you go to the Kwong Wah shopping mall on Paterson Road to buy things, buy hard drives. You have not found it strange, how the price always grows that way, the price is almost fifty hundred dollars. The reason is that many of the stores in the Kwong Wah shopping center, the boss behind the same boss in fact. You think you're comparing prices, but in fact you're only comparing the strength of an air conditioner and how a shopkeeper looks. So you think you have a choice, but in fact you do not.

Technology, innovation, entertainment. All kinds of new and interesting in Baobab friends say. Hey, did you listen to What Dr. Bao's friends say?

Welcome to Dr. Bao Bao's friends. I don't have any friends sitting next to me today, but I have a lot of friends telling me all kinds of new things every day. A few days ago, I had a friend who was actually the director sitting across from me at the recording desk. His friend heard that after listening to our program, he thought it sounded good. He came to apply for a topic. What is this topic? The topic is called Radical market, and the director may not know how difficult it is to talk about. And then this friend probably doesn't know how serious this topic is either. But we do. The point that my friend Bobo made was always that we want to talk about serious topics lightly and lighter topics even more lightly.

Radical Markets in Taiwan:
Extended Reading with Local Perspectives

In fact, this friend is not the only one who mentioned it. In fact, it just so happens that recently there was a campaign on Dr. Bao's Facebook page. What is it? It is the 27th of November, there will be a new activity of the radical market in Taipei. This is a good friend of Yaxin Huang. William Lai, an active campaigner in the blockchain market, has translated many articles and shared this event.

Well, you must be wondering, what exactly is an aggressive market? How do we write it? How do you spell it? Maybe they don't know much about it. Why is this topic difficult? In fact, it is difficult here. I'm holding a book in my hand, and it's a simplified version of this book. It is the simplified Chinese version of the English book translated by the Machinery Industry Press. The book is a short version of the book translated by the Machinery Industry Press. Radical Markets is a new concept that I thought was very, very cool. How to say it? Because for a long time Dr. Bao has been expecting to ask the Lord. That is, in addition to the capitalist democracy that we are enjoying in our society. And this communism, socialism, which we think is no longer very popular and not very smooth, is there any other way. Because we are now living in this capitalist era, sometimes we feel as if we are not so free. Why? Because sometimes the power of capital is not in our hands, it may be in the hands of large consortia, in the hands of the rich, in the hands of people with power. Let's not talk about movies anymore. That is, is there a third way? A new way. Like aliens suddenly gave human intelligence, we suddenly found a new route. The radical market is such a thing, and it's actually very cool. I wrote an article about it on April 1st of this year. I said I was hooked on a concept. This concept is called the radical market. This radical market is like a theory, or like it's like a theory of the opening stone. It is a new direction or a new theory of a new socio-economic structure proposed by two very cool people. The first person is Eric Posner, a renowned professor at the University of Chicago Law School and an international legal expert, a person you may not have heard of. But it is said that his father, this Eric Posner, was called Posner Jr. He also has a name called Big Posner or Old Posner. This Posner, he is the most well-known researcher on the history of American law. It seems that he is also a former justice or former former justice. Posner Sr. is the record holder for the most cited legal research papers in the United States. So his children must be under a lot of pressure and trying to make a breakthrough. Who is this legal expert working with? With a man who is quite a slash. This man is one of my current idols, Glen Weyl. (05:00) He is the youngest professor in the history of the University of Chicago Department of Economics. You know if you have listened to the program you know that we have introduced a school called Chicago. There are many schools of thought in economics. The Chicago School can be a school of its own. So obviously if you can teach in the economics department at the University of Chicago, you must be very good. Why do I say slash? He is not just a full-time economist now. He is also the chief economic researcher at Microsoft, whose stock price has gone up a lot recently. These two if you look up their photos on the Internet are very handsome. These two can be said to be the winning group of people in life, surprisingly in the previous few days, should be last year. They took a chance on the sociological community and came up with a brand new idea. This idea has just saved us from reading the radical market several times.

Radical market, what exactly is it? Before we talk about this, let's give you a feeling. After this thing was proposed, the market was shaken. Everyone to see Amazon book reviews, many people gave them a very positive five-star rating, of course, many people gave one star that can not read. Why, because when I first saw this thing, I felt very surprised. I thought it was great, finally God gave me a response, we finally have a third route. Because everyone used to say that democracy is just a Starbucks coffee. It's not good, but it's not the worst either. If someone comes up with a Blue Bottle, it's very exciting, right? So I really want to study it. I'd like to ask our good friend, Tom, before I buy the book, and Tom warned me that it's best not to buy it. After reading it, you will feel that your IQ is very low and can not read. So he stopped me, what to do? But also very interested, so I started to find some information on the Internet? If you don't find it, you'll be surprised. How do I put it? After the sale of this book, a very widespread movement was started around the world. It is called Radical Market Movement, or Radical Market Exchange. At the same time, the god of the blockchain world, that is, Vitalik, also became his fan. He came with his army to join the Radical Market Movement. And he said that all the engineers in the blockchain world are absolutely dedicated to this idea and are putting all their resources to make this idea come true.

Okay, so what is the radical market? Actually, the bad thing about this is that it's too new. We said it only appeared last year. Even the translation is not sure how to translate the word radical. We say it is the radical of radical, the progressive, or the further progressive. But in fact, some people say that he should actually be translated as basic, progressive. The base of the market. In fact, there is no definite conclusion yet. The simplified Chinese version finally uses the version of radical, radical of radical, progressive. Radical market. It does not matter if you are confused, we just want to tell you the five main points of the latest socio-economic theory in a very simple way today. The five key points, after talking about it is no longer, on the class OK.

First of all, the first point is the self-assessment tax assessment. Let's make it simple. That is, the assessment tax means that you can decide how much your house is worth, and then the society will take 7% of the tax based on the price you decide. She said that seven percent is a magic number, the best number he could calculate. Under this rule, you will not set your house at too high a price. Otherwise you would have to pay too much in taxes. But you also won't pay too much for your house. Why? Because there is a self-assessment tax law. He said, as long as you have assessed the price. Anyone can buy something from you at that price. So if you say your house is only worth $100. To save on taxes, I'll pay taxes on my house for $7 next year. I'm sorry, you may have your house bought tomorrow, and then it will be forced to transfer. But if you say, "I'm afraid this house will be transferred," I say my house is worth $10 billion. You do not have the motivation and ability to do so, why? Because you will have to pay $700 million in taxes next year. Of course, there are some problems that need to be solved here, because there are legal experts here, we will let them solve. He said there may be something that we talk about in terms of the starting alley, or something that has historical significance, something that makes sense for you to talk about this person. It might be possible to avoid being transferred at any time. In short, what he's trying to say is that he wants to get everyone involved in whether or not your assets can be transferred (10:00). The reason is that a lot of people, we are talking about a lot of people talking about hoarding taxes. A lot of people have three houses, five houses, eight houses or 80 houses. But as long as they pay a fixed tax, he can always hold these houses in his hands, he does not need to liquidate these assets, he does not need to transfer it. He doesn't need to move it. He can use the house to make more houses. He pays very little money. The first Radical Markets want to propose to solve, is through the Shared Self-Assessed Tax, his abbreviation is called COST, his full name in English is Common Ownership Self-Assessed Tax. All in all, this I think is not so sexy, but for this theory to speak is very important. This is also why Tom stopped me from reading this book, because there is a lot of mathematics in the book talking about this part.

The second part is a little more interesting. We're going to vote on January 11th, right? The presidential election in Taiwan. Their book, the book by these two authors, proposes a new human voting system through economics and law. This voting system is called square voting, which is cool. What does that mean? In the past, when we were in an election, everyone had only one vote. Then, you vote for the person you want to vote for. But we often find that in modern democratic elections, what often happens is that the least rotten of all the rotten oranges is chosen. Or, in the end, for some reason, the most rotten orange may accidentally get in. Or if you go back to the study of mathematicians, the modern way of voting you can easily in the three candidates to the last elected is that we do not want to be elected that person. Because everyone thinks that I vote a little, the second vote a little. There are some such intermediate stray votes on the third. But then the first two fans, some people may be careless, do not know what reason they feel that he also voted for the third. In short, the current voting system still has its problems. The two people proposed a kind of Quadratic Voting, QV, which he said could bring a kind of protection from the harm of the existing voting system. He said that this one voting scheme could possibly prevent Trump from being elected. What he said, not what I said. The way the square voting law works is that you imagine that you are not given one vote, one vote for one person. Instead, you are given a hundred points. You have to use the points to redeem the votes, and then vote for someone. How do you use these points? If you want to vote for a person, you have to use one point to exchange for a vote. This time, if you think, you really love this person, you want to vote for this person a second time, a second time. Then what to do? You can use two points to exchange another vote, and then vote for this person. If you love him so much that you want to vote for him a third time, you can use four points to exchange for him again. This time you will do the math, the square of ten is equal to one hundred. So the number of times you can vote for the same person is definitely less than ten, right? You can calculate how many times. About seven or eight times. This is when the fun comes into play. She said that according to behavioral economics and mathematics. Most likely, not everyone loves to vote for a candidate so much. It's just that this is normal. Sometimes we elect you say Hillary the Trump. We may not everyone is so love-hate clear, said I will not vote for Trump, I have to use 100 points on him. I have a hundred points to use on Hillary. Or like Taiwan, we have an election this year, right? I heard that Mr. James Soong was going to run for the eightieth time. No, he's kidding. He's probably going to show up again. We may have this, I don't know if President Wang will show up again. We may have this President Tsai Ing-wen ah, we have Mayor Han ah. So, if you use the square voting method, you can even use 100 points on all different people. You can vote for everyone. You can also vote for the person you like the most with 80% of the points, but you finally put some of the few points of the vote to some, you think, for example, like the early Li Ao out of the time, may be some people will feel very supportive of such a third force. This time, it is possible to let more of us talk about the new candidates of the non-vested market to get some new possibilities. He said, according to mathematical calculations, this third choice, and is a good choice, easier to emerge in this voting method.

The third one is called the unity of the workers of the world. What is this? He offers a wild idea called the immigrant bonus system. What does that mean? We all know that talents have to move. If Taiwan wants to develop, we may need not only people to go out, but also talents to come in. What is the situation in the United States now? Localism. So they want less green cards to be issued. It's harder to immigrate, you really have to be very, very good to get in. But this may result in the ticket being in the hands of a few people. For example, (15:00) the royalty will be able to immigrate smoothly, to be able to become the new American talent and so on. So they came up with this method which is cool. He said that any person you can go to introduce a foreigner you know. If you think this person is very good, you should be able to vouch for him to the government and then act as his visa introducer. If you successfully introduce a very talented person to, say, Taiwan. One percent or two percent of his next job will be yours. It's a kind of referral fee concept. He wants to use the logic of referral fees for the flow of talent in a country. So he may try very hard to help your country, to introduce great people to work. If you introduce the wrong people, then you will earn very little money, and there may even be a fine I don't know. Maybe he will have some design afterwards. For example, you may need three years in a row to receive the first payment. So it's useless if you refer randomly. If you do introduce the right people, several. You may even be able to make money by introducing people to Taiwan to work and set up companies. All in all, it is called the world has the ability to work people, should be united. His meaning is actually this. But because his name speaks a little communist, so it is a little criticized by everyone, you are selling people as commodities? This kind of concept.

The fourth one is also more serious. It is about monopoly. The first one is a monopoly of the market, which has been in existence for a long time. The company's main business is to promote the development of a new product, the octopus. He said they went to do a study and found that the top six shareholders of the United States banks are the same several large consortia. For example, BlackRock these large consortia are jointly held. That is to say, the banks may appear to be different families. But then, the board of directors behind the board, many of the holders may be the same people in possession. So they will probably join together to do something like interest rate manipulation, or the simultaneous raising or not raising of commodities. So this may affect the market opportunities, and the possibility of our choice of commodities. A little like, I don't know if you know, you go to the Kwong Wah Shopping Center on Paterson Road to buy something, buy a hard drive. You have not found it strange, how the price is always the same, the price is almost fifty hundred dollars. The reason is that many of the stores in the Kwong Wah shopping center, the boss behind the same boss in fact. You think you're comparing prices, but in fact you're only comparing the strength of an air conditioner and how a shopkeeper looks. So you think you have a choice, but in fact you do not.

So, she said they found that there are a lot of such things in the capital market now. You feel like you're choosing between different builders; you feel like you're choosing between different banks; you feel like you're choosing between different commodities; you feel like you're choosing between different political figures. But in fact, you may choose the same person, the same group of people, the same organization. At this time, if they are good to you, you are good. If they treat you badly, you are bad. You simply have no other possibilities. Okay, so this is what they want to prohibit by decree. A large consortium can not invest in more than two companies of the same type. That is, if you really want to invest in banks. Please, you can only choose one to invest. You can not invest in several banks at the same time, and then you may create a so-called joint this action.

Well, we should probably still have some spirit now. Let's talk about the last one, which I personally like the most and I think is the coolest. It's called "Data as Labor", what does it mean? He said that recently, there are rumors in this society that AI will make us lose our jobs. We humans will have no jobs to do in the future. So what happens if you don't have a job, you don't have an income. No income, no money. Without money, there may be no way to live. They say, according to their economics calculations. The chances of this happening are very, very high. So he thinks we should come up with some solutions in advance. The AI's teacher is actually us, right? We know that AlphaGo is very powerful, and the teacher of AlphaGo is his teacher, that is, the Go games played by all the players in the past 3,000 years. So, if Google has an AI, Facebook has an AI. why are they so good? How do they know what the answer to a certain thing is? Actually, we gave it to him. And when did we give it to him? Have you ever thought about when you usually slide your phone, you sometimes click a like, click a like. You go searching for an article, you click on the ten options inside, you think the best and the best one. In fact, we slide the phone, thumbs every time the slide process. Every time we press this click, we are teaching a super smart AI (20:00) how we feel and what we think is right or wrong in the world. And then, what happens? These AIs will grow big. Thirty years from now, they may replace my job. It is possible that there is a super

Radical Markets in Taiwan:
Extended Reading with Local Perspectives

powerful AI behind YouTube, and one day it will be able to simulate a Dr. Bao and teach a lesson exactly like the one he taught, and we won't be able to tell. At this time, we may still choose the fake Dr. Bao when we choose the online course to purchase. But the person who gave him the ability to do that was actually me. I uploaded the video, I slid the phone, everyone clicked on it, everyone shared it. So what is the meaning of data as labor? He said that these companies should be regulated, and in 30 years, each person should be paid 5,000 dollars a month. To let our previous thumbs in the slide phone labor can be distributed to us. It is his future earnings may be billions, he should want to share some to us right. The process of growing up is fed by our thumbs. How can you take away our jobs and then make it impossible for us to live? He said that data is labor, so I think this thing is very cool. I think this is a very cool thing. in fact, to put it bluntly, you are now slipping your cell phone every day, so that the enterprise collects information, it is also helping him train AI. if thirty years later, these AI really replace our work. Each of us should get a little data labor subsidy every month. And this labor is what we now slide the phone to bring. In fact, this is a practice that I feel very supportive of. The author Glen Weyl once deep inside the base camp Google speech, it was attacked by many employees' questions. The company's main business is to provide a wide range of services and services to the public. We do AI. first we have not made any money yet, right. You see AlphaGo now also can not make any money. The second, we have not given you feedback? Yes. You use Gmail every day you have to pay money? No. Do you pay for watching YouTube? No. These cloud hosts are very expensive. Every year, billions and billions of dollars are spent in this way, do you pay? No. So I'm like letting you earn five dollars a month in Gmail subscription fees, right? Every month you have to pay the $249 monthly fee to YouTube, I let you not pay. So it's like you're feeding us this information, your thumb's work, and I'm giving it back to you now. But Glen Weyl is an economist in his own right. He calculated that it is possible that we are even now. But you give him another 15 years, give him another 30 years. The same AI can earn the number of money, will be exponential growth. It may be thousands of times, tens of thousands of times or even hundreds of millions of times more than it is now. And then from the reduction of our income, or the number of jobs that may be taken away, may also be several hundred times more. So he took the results of this calculation and spoke for us. Just say, not just now let us use Gmail for free on it, not now let us watch YouTube for free on it. You should formulate a standard earlier, so that the day we cry to the sky and say we are poor, these large companies should transfer a little money to us every month. It sounds like we are poor, but don't forget that we may be really poor in the future. The company's main goal is to provide a platform for the company's customers to learn more about the company's products and services.

All in all, this book introduces these five very important new ideas that they have come up with in considerable length. Then, in fact, during this period of time, the whole world has doubts about the status quo, just like Dr. Bao, we think that besides Starbucks, besides Louisa, do we not have a third choice? Of course, there may be Seven-Eleven coffee, I also often buy. But is there anything new? We seem to think that we are on the side of justice and the happiest utopia. A capitalist liberal society. But why do we feel like we don't have this utopian feeling? Why do we still feel that we are not paid very well? Why do we still feel that we can't afford to buy many houses? Why is it that rich people can borrow 99% of the money to buy a plane? When we go to buy a house, we desperately ask for love and say can I get a bigger loan? At the end of the day, you have to put up a 10% or 20% of your own money. All in all, this radical market, hopefully through the academic perspective. They really are super sacrificial. I must say, after they put forward these proposals (25:00), they have been criticized in the academic community and in the industrial sector. Many people supported him, but many people criticized them. The first criticism is that you are selling people as slave labor. We just talked about this is called the visa bonus system. It's a little bit like we are treating people as commodities. And then some people criticize that the self-assessment tax system is a communist system. How can we own our own property when our house can be bought at any time? What's so different about this common property? Many people have criticized this. However, he actually gave a speech on Google, which I highly recommend you to listen to. He said that he proposed this theory to be the most radical capitalism and the most radical socialism. He also put two big heads, one called Adam Smith, one called Marx. He said that my theory was a new line between the craziest Adam Smith and the craziest Marxist advocates. So Marx would not agree with him, and neither would Adam Smith. It was something completely new. One of the coolest things about it is that he felt that the direction of this idea had actually been tried by a man decades before. This man is called Sun Yat-sen. In his speech at Google, he really put a picture of Premier Sun Wen. He said, Sun Yat-sen back then there is a called average land rights, and then there is a government can buy back ah. He said that the design was very much inspired by George Henry. In fact, there was a direction, but it was not fully implemented. So, he felt that he had the responsibility to think this thing through and then add a lot of new things to it. Including our current voting method and so on. Many people support this idea, including Vitalik Buterin, the blockchain god, who mysteriously showed up

Radical Markets in Taiwan:
Extended Reading with Local Perspectives

in Detroit on March 22nd to participate in the first global conference of the radical market. It's called RadicalxChange. x is x. Change is change. So you can also take xChange as a kind of exchange, a kind of I think we should have a new flow in this world. You can also take Change as a kind of change. So in Chinese, we turn it into a movement and organization for radical change. RadicalxChange chapters have been established in dozens of cities around the world. This year, RadicalxChange Taipei was founded in Taipei by some of the very cool people we've just talked about, such as Yashin Huang and Chen** (28:02), who are all working hard to promote this idea. In fact, there is another very powerful promoter in Taiwan, this person is called Audrey Tang. Audrey Tang, in fact, also this year smoothly, it should be reasonable to become a member of the chief committee of the global organization Radical Markets RadicalxChange this movement. So, there will be some such movement in Taiwan next. Or, let's not talk about the movement, the movement is too serious, is some new changes.

Then you may think, "Didn't Dr. Bao talk about blockchain? In fact, the blockchain technology can really realize many of the ideas proposed by the radical market. For example, let's talk about dividends. The virtual currency of blockchain is very convenient to do the so-called dividends. We talk about data as labor, that is, I want to play a contract with the government and enterprises, after 30 years I will get a sum of money, a monthly sum. This is very suitable for the use of intelligent contracts to implement. Or you can say that there are a lot of restrictions in it, including the support of the big octopus you can not have corporate monopoly. All these are also in line with the so-called transparent and open approach of blockchain. In fact, the spiritual leader of the blockchain world, the prince of law, Posner, and the economist's star of tomorrow, in fact, he is already the star of today, Glen Weyl, have combined to create the possibility of a new social system for the future of mankind, and we should not say that it will definitely succeed. That such a thing is actually very, very attractive Dr. Bao. So, you are also very encouraged to search more on the Internet if you have spare time. You can search the radical market, you should find my article. Or a lot of articles translated by Taiwan, or various people. (30:00) Or, you can choose to attend the second or third RadicalxChange Taipei on November 27th at the MaiCoin Building, which is also a great Taiwanese company in the blockchain, at the intersection of Bade Road and Xinsheng South Road.

Well, I know that today's content is still rather serious, so there are no jokes. Glen Weyl, in his speech, proposed a word called marketopia, a market utopia. What he meant was that we economists used to hope for a free market without too much government control. Too much government control is totalitarianism. So we should allow the market to operate freely. We think our current capitalism is a free market mechanism. But in fact, if you look carefully, we have just said that the market is really free? Is this market really in our hands? Is the capital of capitalism really on our side? The market is really free. The real marketeers can think that this is like a utopian world. Will such a world come or not? Or whether there are enough people who think that there is still room for progress in our current society, it is difficult to say. Many people have been drinking good Louisa coffee, good Seven-Eleven coffee, and good Starbucks coffee all their lives, so they may think that there will be no more Blue Bottle and no more Blue Bottle coffee in this world. But now we have a third wave of coffee revolution, why not in the market or world of our socio-economic system, there can't be a third wave of revolution?

Well thanks for listening to today's blogger friend, I'm Dr. Ruthie Ge. If you like my show, remember to subscribe to my show on SoundOn. If you are listening to this program on other platforms, please help me to give me five stars by leaving a comment or click the little bell to follow the subscription. See you on the air next time, bye bye.

Note: Some of the text in this chapter is marked as time code due to dictation problems, so please go to the original release channel for further information.

Radical Markets in Taiwan:
Extended Reading with Local Perspectives

Radical Markets in Taiwan:
Extended Reading with Local Perspectives

9. BBFT PODCAST #EP10 *"SOMEONE SLIPPED THE PHONE AND BECAME THE UNEMPLOYMENT CRISIS FOR EVERYONE AFTER TEN YEARS!"* FT. DR. SUJI YAN

Originally on SoundOn & YouTube
https://youtu.be/NUWKK9jsmOs

Dr. Bao: What does your old driver mean?

Suji: Is it the original Yunan mountain song?

Dr. Bao: But now the old driver does not mean something else?

Suji: Is that what he meant originally?

Dr. Bao: In the lecture about people seeking flowers.

Suji: Yes, the original song I will not sing, on what the old driver to take me, I want to go to what what, I want to go to Yunnan, there is a music video is very exaggerated.

Technology, innovation, entertainment. All kinds of new and interesting in Baobab friends say. Hey, did you listen to Dr. Bao's friends say?

Dr. Bao: Welcome to our friends at Baobao, I am Dr. Ju-Chun Ko. In the fifth episode, we talked about the five key points of the radical market. The first self-assessment tax, the second square voting method, the third immigration dividend system, the fourth monopoly octopus, the fifth is my own favorite data that is labor dataaslabor. five key points you still remember? The fifth is my own favorite data that is labor dataaslabor. It is said that a lot of people find it quite

Radical Markets in Taiwan:
Extended Reading with Local Perspectives

simple and easy to get started. Today we will invite to the radical market sector of the high people, in fact, he is not radical market sector of high people. He is a senior person from all walks of life, he is a senior person from many walks of life to our program. Let's welcome Suji Yan.

Suji: Hello Hello everyone, I am Suji. because someone said turn over and read is that eat that thing.

Dr. Bao: Salted crispy chicken, crispy chicken.

Suji: That doesn't really matter.

Dr. Bao: I always wanted to say whether the driver had anything to do with it. So I just asked him a slang term, the slang term for an old driver.

Suji: Yes. The old driver is from Yunnan.

Dr. Bao: Yes. Everyone go to Google. clouds empty a veteran driver, you will know what we are talking about.

Suji: Suji it is my original pen name. He was like this, he wrote the first word in Chinese. He can be pronounced by Japanese people. Then, it seems that Americans will also pronounce it, which is good. The earliest is in Q daily curiosity daily, there are some articles written about self software. I will use this.

Dr. Bao: Kind of like a pen name.

Suji: Later anyway, when I start writing things, I will mix the two. If it's a prefer with English name, or with that kind of pen name, then I will use that. Then he must use the real name, I will also use the real name.

Dr. Bao: Is your real name not to be revealed?

Suji: No, you can show it, just look up Suji Yan, and then you may see the real name.

Dr. Bao: Well, Suji, it's kind of like Dr. Bao is a stage name. My real name is Ge Rujun. But I really think it's interesting when you just said that. It's not that easy to find a name that over 80% of the world's population can pronounce.

Suji: Because that pronounciation is very difficult is something.

Dr. Bao: Mine, like Jun, is turned into ** (02:51) and it's hard to read like that. When I went to Japan, I was very hesitant about what to call it. I was very hesitant when I went to Japan about what to call JC, because if you say JC, Japanese people don't pronounce it well, because his name will give you a san, right?

Radical Markets in Taiwan:
Extended Reading with Local Perspectives

Suji: It's like a funny man.

Dr. Bao: Yes.

Suji: Yes, yes, yes. Look at that PICO, PICO Taro, two short syllables of letters, it will look like a funny artist.

Dr. Bao: So at that time they felt a little unnatural expression. So I later had to change my surname, because I found that many Taiwanese people go to Japan, everyone is called surname. But at that time I didn't use sex, just Ko, Ko San, because many surnames in Japan are called Ko San. The surname Hsu is also Ko San, the surname Jiang is also Ko San, Ge is also Ko San. So all Taiwanese people are called Ko San. I really used JCKo for a while and failed. I've been using JCKo for a while, but it's not working.

Suji: Hello hello hello.

Dr. Bao: Just go with it. Okay, today's topic is not me, or on Suji. Let me tell you about Suji's very interesting experience. He is now the founder and CEO of Dimension.im, a new start-up company. He had a very interesting school experience in the past, and we'll ask him about it later. He went to the University of Illinois at Urbana-Champaign, which is the most prestigious school in the United States. And he graduated with a degree in Information Engineering. But fortunately or unfortunately, he didn't graduate, not yet. Maybe he will have a chance to be like Steve Jobs later.

Suji: He graduated?

Dr. Bao: go back to get a card, honorary. Then, the former Qdaily, Qdaily on Curiosity Daily, the independent reporter of the financial media. And then former engineer at TuSimple, a self-driving company. Then it is also this radical market RadicalxChange, Data Law Group and other economists, legal scholars, with many people to collaborate on writing articles, written on the Internet, I have read. A while ago he did an action (05:00) that was reported by many international media. Including the world's most famous online magazine, it has since become wired originally magazines, but now online has been very famous, "South China Morning Post", "Wall Street Journal" and so on to do coverage. His current personal is focused on open source, encryption, privacy protection. There is one, on Dimension is the company, then you have a product called Maskbook. we now torture to ask, Dimension is we talk about how a dimensional dimension? Tell us what you are doing in this new company?

Suji can actually go to the official website to see, it has been written, I remember the original called Cyberian life matters.

Dr. Bao: Translate it.

Suji we all call ourselves citizen.

Dr. Bao: Citizen Citizen.

Suji Citizen, there is a very famous movie called "Citizen Kane", "Citizen Kane".

Dr. Bao: "Citizen Kane.

Suji right, speaking of the 1900 to 1950, that big time inside there is this market development and depression inside. Giant monopoly and government power and human nature between the alienation of these relationships, this is how I understand.

Dr. Bao: When is the movie?

Suji black and white film.

Dr. Bao: It sounds quite avant-garde.

Suji black-and-white movie, the main character called Kane. then, he suddenly inherited a super family estate from childhood. Because his family that land dug out the oil, and later began to do newspaper.

Dr. Bao: So in telling his story.

Suji to, called CitizenKane.

Dr. Bao: Is it a true story or a script?

Suji had a prototype, a newspaper tycoon, who later had a palace built and then died. But this movie is very good. Our general identity in this real society. Our identity on earth, in every country is a citizen, a citizen. The term existed in the 1980s, and since the 1990s, mainly Tim Berners Lee, who invented the World Wide Web, there is Netscape Navigator browser, after which there is a term called netizen.

Dr. Bao: Netizen by net and citizen.

Suji has different translations, there are those who are netizens and those who are netizens. But netizen is a very soft translation. It's a particularly cynical translation. Netizen also gives the right to this word. His idea is relatively simple, is to have a network, we go there to play, is this feeling. The latter has a school of science fiction or, including the movement of some people, he will use another word, called cypher or cyber.

Dr. Bao: Yes, I actually have some points not quite come out, this cyber cypher seems to be interoperable is it?

Suji is not, cyber is, cyberpunk is a type, a genre. It's talking about that kind.

Dr. Bao: Cybe rspace.

Cypher punk is about that one particular movement. It's a movement that uses cryptography to do encryption.

Dr. Bao: Cypher comparison is biased.

Suji only has a password.

Dr. Bao: Only the cryptography.

Suji Password Punk, Crypto Punk.

Dr. Bao: Cyber punk inside the cryptography of these people, we can say he is Cypher punk, but there are some Cyber punk may not write programs or do not engage in cryptography.

Suji right, but they have a lot of generic for the future of some vision, that is, whether it is fear or expectation, there is "Blade Runner", I remember the translation called "Silver Wing Killer". Blade Runner" inside the most famous is the artificial man died when that sentence, right. There is also "Ghost In The Shell".

Dr. Bao: I'm very moved now that I think of that.

Suji I can not recite.

Dr. Bao: See what a certain light on Titan, what a wow, that. That actor recently.

Suji he passed away.

Dr. Bao: So he's gone to Titan.

Suji right, then, there is the more famous Japanese.

Dr. Bao: "Attack of the Shell Motor Team".

Suji right, Shiro authentic, called "Ghost In The Shell", it will take Hong Kong or to this kind of East Asia, it is not necessarily East Asia, that is, the population is super dense, mixed race, speak messy language.

Dr. Bao: Super City.

Suji right, and then everyone there with strange equipment. There are all kinds of lasers in the sky that feel like a city. This art form of describing the future, whether it's a novel or a comic or a movie, they're called cyberpunk, that Cypher punk gang, or all kinds of people, they may have a vision of the future is that way, in the eighties, nineties, think the future may be that way. The World Wide

Web appeared in 1990.

Dr. Bao: It is called the Internet in Taiwan.

Suji Internet, more boring. It is more boring than that.

Dr. Bao: Because there is no picture, right?

Suji no picture, and you say that artificial intelligence he does not have a beautiful woman to go with you to do what you do.

Dr. Bao: There is nothing floating in the sky either.

Suji right, very boring. But then, there is a word called Cyberian, that is, in the Cyberspace citizen, Cyber citizen, it is actually greater than the meaning of the netizen.

Dr. Bao: So you mean your company name at the beginning is there to want this.

Suji right, because it is another dimension well.

Dr. Bao: You just spoke too fast, you speak again.

SujiCyberian life matters, is one of our, the official website has written this, right.

Dr. Bao: Tagline, tagline.(10:00)

Suji that this is to learn that many movements it will have what so and so life matters, Cyberian is to say refers to the citizens of Cyber society. It is a concept that transcends age and geography and sovereignty. That is we may all be a cyber space citizen. then we will use this word, right. We have written underneath that we want to provide a series of tools to empower, some citizen like this.

Dr. Bao: This is so cool and romantic. I don't know how people might feel about that. So I guess I'll start by asking, "Does AI AI count under this category? It is virtual life covered under this?

Suji, it is possible that he is also a Saibo citizen.

Dr. Bao: You are open to it. You are open to definition.

Suji I was thinking is.

Dr. Bao: Have you ever seen a movie called "Her"?

Suji right, and that "mechanical Hime", "Ex Machina".

Dr. Bao: That like this sub called Her it is considered this citizen?

Suji, I think it doesn't depend on our definition anymore. It depends on the citizen of the time, how acceptable he was to society? Yes.

Radical Markets in Taiwan:
Extended Reading with Local Perspectives

Dr. Bao: I don't know how we feel about this, but the density of information in a very short period of time is very huge. I myself feel how only three minutes after the beginning, I have not even warmed up, my brain began to heat up. This is cool, especially cool. Say it, you have done so much, you pick a few points to talk about. I speak one by one in order, I see that our program today may record two hours recorded not finished. You first talk about, is your this school, because in fact on my understanding, I first encountered Suji, should be the Flying Tiger introduced?

Suji vs.

Dr. Bao: We were shooting at that time.

Suji photographed the scene, right.

Dr. Bao: We were engaged in the "Saints and Thieves" at that time.

Suji photo shoot live.

Dr. Bao: My audience is very annoyed, every episode to talk about these four words annoyed to death. We are working on a blockchain project. At the beginning. I often tell them that the blockchain project is a lot of fraud, that is, not serious. To be serious you have to take at least one picture of everyone in the same style.

Suji is right, you can't search directly.

Dr. Bao: Yes, you can't just find a posting and the picture quality is different, the background is not the same, the posture is not the same. What do you call that? This team is probably far away from 80 different countries that kind of feeling.

Suji and neither exists.

Dr. Bao: are likely to not exist at all, it is likely that E-mail simply did not return the letter put him on such. So we took a picture, Feihu then told me, introduce you to an interesting person to know so. And you came. That time your hair style is like this? You seem to have a little different look at that time.

Suji I have long hair.

Dr. Bao: Yes, I remember you were pretty cool at that time. He said this person is special, from the East University, that time you are still East University or?

Suji did not, I have come out.

Dr. Bao: has come out, feel the feeling of being out of prison, a feeling of being out of prison.

Suji has a little bit, I grew up in Shanghai when I was a child. My parents are also from Shanghai. Then, after my senior year of high school, I went to the United States about 17 or 18 years ago. It just so happens that I didn't end up taking the college entrance exam, so I went straight to study information and computer engineering.

Dr. Bao: You went straight to the exam?

Suji Yes, in fact, there are many people, but perhaps, I have counted quite a few at that time, and later will be more.

Dr. Bao: So you are also the one who scored full marks?

Suji has no no.

Dr. Bao: I heard that Mark Zuckerberg is a full 1600.

Suji didn't, I should have.

Dr. Bao: Well, you can stop talking.

Suji that time has been changed, to 2400 points.

Dr. Bao: But it's not easy to get this exam.

Suji is actually not too bad, because there are other schools that accept students. But they have the highest tuition cost ratio. Public schools are cheaper.

Dr. Bao: Anyway, there you go.

Suji later found out that it seems that the university is not quite the same as I thought.

Dr. Bao: You originally thought it was a cyber-citizen's paradise like this, and then everyone in it absorbed the knowledge of the Bible like this then?

Suji didn't. I found that everyone wanted to get a job.

Dr. Bao: Do we all want to find a job or get a job?

Suji is looking for a job.

Dr. Bao: Looking for a job?

Suji Yes, of course I'm not saying it's not good to get a job, I want to get a job too, it's good, it's cool. And then, Facebook and Google will come to recruit people. But I think it's not that interesting because I have a better relationship with many professors. I asked him if it was like this 20 years ago. He said it started slowly from a certain year to become this way. You have Marc Andreessen, who came out of UIUC in '94. And then there was the founder of YouTube.

Dr. Bao: So Marc Andreessen is the University of Champagne?

Suji vs.

Dr. Bao: We've talked about this man about a hundred times in the program.

Suji he is very interesting.

Dr. Bao: Who else do you think?

Founder of SujiYouTube.

Dr. Bao: The founder of YouTube, Shijun Chen.

Suji, his Taiwanese counterpart, Chen Shijun, who recently moved back to Taiwan, is also, he should not have graduated, he should be one semester short of graduation. The later entrepreneurship is becoming less and less, not entrepreneurship is becoming less and less, accurately speaking is to join the entrepreneurial company, join the big company, join this very fashionable company people are coming more and more.

Dr. Bao: The ratio has become mainstream.

Suji but if an anti-mainstream thing becomes mainstream, then the anti-mainstream people become less.

Dr. Bao: So finding a job at that time was mainstream?

Suji is right, you are right to say so. Everyone is doing things, and the things he's doing are not particularly interesting. (15:00) In fact, there are also interesting, maybe I know less.

Dr. Bao: just feel no longer interesting.

Suji, I spent my freshman and sophomore years at UIUC. In my third year, there was a program with the University of Tokyo for the first year, so I could stay there for one year or two years.

Dr. Bao: So good?

Suji right, he will be some people have degree it I remember is. But the main thing is still this to visting as the main, visting student as the main such feeling, will play well.

Dr. Bao: So that's good.

Suji I suddenly found that is, he does not seem to want money again.

Dr. Bao: Another function of the University has suddenly appeared.

Suji is right, he will send you money, I learned to speak Japanese.

Dr. Bao: You would have.

Suji I already know, I learned it in high school. But maybe I don't speak it particularly well, I meet some basic requirements. I found out that the Japanese government gives you money, and you don't have to pay the tuition in the US. Anyway, it's very cheap, so I'm going to go.

Dr. Bao: It's not a matter of price/performance ratio, it's completely.

Suji send money, right?

Dr. Bao: Yes.

Suji I went before, because it is not the same academic structure, one is the start of school in April.

Dr. Bao: Yes, that's much worse.

Suji a month is the start of school in September, the middle will be a little right. Not right, because you are in the middle of the school out, so you can apply for that CPT, is in the case of no visa can go to work. That is, because it is possible. In Japan and the United States can be.

Dr. Bao: All can work part-time?

Suji works part-time.

Dr. Bao: Follow up work.

Suji Entrepreneurship doesn't seem to work.

Dr. Bao: Come work for the company.

You can also work for Suji, so you can go to work full time.

Dr. Bao: So you're just a few months away?

Suji seems to be almost a year.

Dr. Bao: Almost a year?

Suji 15 years of several months forget, to 16 years of March.

Dr. Bao: So you didn't attend school during that time? You didn't go to school because you deliberately put it.

Suji no, it is originally this way.

Dr. Bao: This is the way it is, not intentionally?

Suji it academic system is not the same.

Dr. Bao: So you have an opportunity to go to work?

Suji right, to the middle of that gap is relatively strange, no one cares about you. Maybe, he also does not charge you money, no one cares about you. You want to do work, you pay the government's taxes on it, to the U.S. government to pay taxes on it.

Dr. Bao: So you're just a ronin in Japan like this?

Suji is not available in the United States.

Dr. Bao: Cyber wannabe.

Suji is in the United States.

Dr. Bao: Being a ronin in America?

Suji was in California and went to the company called TuSimple.

Dr. Bao: So you explained to, this we speak TuSimple.

Suji because it is very strange, why you did not graduate to do what the drones. The company was very new at the time. The CTO was also very good, and they were coming out of *** (17:25). He suddenly wanted to recruit people at a strange time.

Dr. Bao: And then you go?

Suji no, normal people will not go.

Dr. Bao: I still think you after you do luggage packed, you flew to Tokyo, on lost in translation.

Suji did not. It's also quite strange because of the timing of its recruitment. I have forgotten exactly, but in my impression, there should be no one to work full-time at that time, no matter it is intern or full time.

Dr. Bao: Maybe Christmas this logic.

Suji because they are entrepreneurial companies, I can not say I wait until after the New Year what to recruit people, he just recruit, I just interview, now that the company is still quite famous, has been a unicorn

Dr. Bao: Now it's a unicorn?

Suji right, now on unicorns, on three years.

Dr. Bao: I can't read it yet, I just asked about it.

Suji this is a famous word, everyone look it up, TuSimple. no, is a famous leader said.

Dr. Bao: I think you continue to talk, we may have to record this episode for 3 hours.

Suji later I was at that place because I had also taken some machine learning classes and AI classes. I took some junior and senior classes in the second year. I really like this piece of AI, so I finished that, this strange time recruiting people will not come, you will come to work, then I went to work there.

Dr. Bao: But essentially you are still a coder, right?

Suji right, I am coder, I am on that side to do a platform with data and AI related. then collect data, put that data how, to put it algorithm experiment out.

Dr. Bao: So you went to Japan in the end?

Suji went.

Dr. Bao: Where are we in the chronology now?

Suji was there in April of 2016, when school started. I stayed until the end of 2016, I think it was September or October.

Dr. Bao: That's quite long, you still endure a lot of time, you said you do not like the school? You do not want to enroll.

Suji anyway, a total of three years, two years in the United States, less than a year in Japan, and then back to deal with a little thing.

Dr. Bao: Something happened?

Suji no, I deal with that dropout thing too.

Dr. Bao: Oh, to deal with a little thing. ok, so in Japan is in the East University on the stay, and then did a little research?

Suji right, then it is so, because at that time there are some media and they have a very good relationship, do not know why, is to know the network friends. Then, they wrote very badly, and not very badly, sometimes I think this actually I can write better. Because I originally wrote something, messy things. Chinese media. I'll talk to them to let me write something? They said yes, because I freelancer. then I have a visa in various places, and then he also think it does not matter. He said, then you want to write what you tell me, you need funding we give you approval.

Dr. Bao: You started writing?

Suji vs.

Dr. Bao: You are quite good at writing, I see you three days ago also posted a.

Suji Radical Market Book Review.

Dr. Bao: Is the book review of Radical Market complete, or is it just about a particular chapter?

Suji is complete.

Dr. Bao: Because it was too long, I slid for a long time.

Suji is complete, and that is still cut off, the total original text should be 6000 words.

Dr. Bao: And where is the original text?

Suji's original article was later issued again because it did not have enough space in the newspaper.

Dr. Bao: Seems to be placed in a financial what?

Suji "Economic Observer", but later on the web page is enough. So the original article is released.

Dr. Bao: So I slide that very long is long.

Suji is the web page.

Dr. Bao: But the entity has.

Suji has a newspaper, there should be printed a little, but that page is very small should be.

Dr. Bao: Just a little bit of deletion.

Suji is right, because it has to be stuffed into the printing space.

Dr. Bao: You may have to use augmented reality to see it, so you're over there and you've written an article in Japan.

I wrote a lot about free software while Suji was in the US and Japan.

Dr. Bao: So you don't need to do research?

Suji is not, because I have a bachelor's degree, so no one cares about me if I don't do well.

Dr. Bao: Free time.

Suji he will give you money.

Dr. Bao: Who gave it to you? The laboratory? The laboratory gives you money.

Suji Japan. No, in fact, do not have to go to the research laboratory, you do not matter if you go to class, it does not matter if you go. Because the University of Tokyo, I do not know what he is doing.

Dr. Bao: special cattle, in short, you people there they recognized so. A kind of gas, your aura, aura in that they on the line.

Suji is right, so happy ah. That time is still better, but you still have to go to report, because he wants you to go every month to do what to do after the money to you.

Dr. Bao: What do you call a "fucker"? Signed?

Suji signature and a general description of how it is.

Dr. Bao: to ensure your personal safety, hands, arms not missing one, so give the money. Anyway, you're over there, I hear, and I know you have a legend that you've even been on a Japanese show? You were there as a funny artist.

Suji forgot the name. No, it's not. There's a **** (21:43) at Eastern University that's very famous.

Dr. Bao: What is that thing?

Suji Dongda's base, you see Japanese anime often have the kind of what invites people to play and do barbecue.

Dr. Bao: East Grand Festival.

Suji he will reporters to come over to shoot, I do not know why they shot me, but also invited me to go, anyway, on the talk about ten minutes. But the appearance of the time may be relatively short, appeared on, I did not record, but appeared for a while.

Dr. Bao: By that time, you already had a long hair style.

Suji has long hair and short hair, then later you will find that staying up late will lose hair.

Dr. Bao: No, it's actually quite good. There is a video today, I think it's okay. My hair may still be less than yours.

Suji has no no.

Dr. Bao: I think it's actually quite interesting to say that powerful people will introduce powerful people to people who are not necessarily that powerful. I am that not so powerful, flying tiger we just talked about this person several times. It is Tang Feihu, he seems to be a legendary figure, we have the opportunity to find him again. He what China programmed the beauty of the contest national champion of this logic.

Suji on that is ACM.

Dr. Bao: Yes, is also playing this kind of mathematical and scientific competition, what to play the mathematical Olympiad such. It was a genius among geniuses. Flying Tiger said that you must know this person, and at that time Suji was ripped out of the chisel, and we got to know each other. In fact, we didn't interact much with each other during the process, we were working on our own (23:03) block chain, and you were working on the Dimension chain, and there were some plans at that time. And now we're suddenly having more interactions, in fact, we're talking about this topic today.

Suji, because at that time I seem to have told you something about social networking. But it seems that Glen's book was not out at that time.

Dr. Bao: Glen is the main author of Raical Markets, a divine book in my mind.

Suji in fact he wrote very well, I was not only with Dr. Bao, but also with other friends in Taiwan to talk about this, on I think it seems a bit problematic.

Dr. Bao: What do you mean there is a problem?

Suji is not, I was thinking that there was a problem with the social network, I think I told you that.

Dr. Bao: Did something happen to Cambridge Gate at that time?

Suji is out, but not yet in Congress. Not so often to get it over to Congress. And then, as I said, the social networking problem, I want to do something to change it.

Dr. Bao: In fact, you are already a Cyber citizen.

Suji Saibo Citizen.

Dr. Bao: CyberCitizen has been around for a long time, and you've done a lot of things, writing articles, software programming, and free software open source, so you're very sensitive to it. At that time, you are not only aware of something wrong, but you already feel that the problem is going to emerge.

Suji, in fact, at that time, it actually did not have some very strong theoretical basis, or in fact, many people probably still in a relatively hazy state.

Dr. Bao: I think this hazy, talk about this still sounds good, I think that is called freedom bubble.

Suji on, free foam.

Dr. Bao: I think we are living in that bubble, we feel that we are very free, want to do what to do, right? I want to upload a photo, can ah, can upload, right. What I want, how I feel, nobody cares about me. Really? Now we all suddenly find that things are not so good, sometimes a post will disappear. So at that time, in 2018 you felt this?

Suji is actually accurate, (25:00) I came out to start a business at the end of 17, the third quarter of 17 years, encountered some investors, they think you are good people, also no matter what you do, we slightly invest in you a little.

Dr. Bao: In Japan?

Suji is in the United States.

Dr. Bao: You were already working with blockchain stuff at that time?

Neither did Suji. I was studying various social networks, and there were minor problems with various social networks themselves.

Dr. Bao: Cyber World?

Suji was right, the social mapping and social capital of the cyber world. But at that time, the blockchain itself was quite chaotic, and I think we all didn't know what to do.

Dr. Bao: So you just get it too?

Suji right, so in fact our past investors are financed by the equity.

Dr. Bao: So it means that there has been no coin to chisel investors, let's not talk about that. We've always used the general company investment approach to find some investors.

Suji, later we were in, in fact, at the end of 17 I went back to China, back to Shanghai. Just sit in the office, on the official have a small team, also want not in other countries also engage in office, anyway, mainly or in Shanghai as the main.

Dr. Bao: That's cool, we'll get him again next time we have an episode. We'll talk to him when he comes to Taiwan once in a while. We talked about data as labor last time, but I don't think I'm deep enough. So maybe we can ask you some more. The other one is the product, Maskbook, and we can take a three-second break to listen to some music today because our brains are tired. Okay, after the break, let's talk about it, because the radical market has actually strung together many of us from different backgrounds. Including Feihu is also very concerned, I am also very concerned, Vitalik, and then economist my good friend Tom, also kind of let us interact also began to more up, I also went to listen to your speech a few days ago. So first of all, what do you think the radical market has to do with us people?

Suji related. Because in fact I wrote a book review for a mainland, official media, that is, the Economic Observer, about the radical market. I began with them saying you should not write fiercely, I said I would write directly out. At the beginning, I wrote about two things, the 1991 Soviet Union, the collapse of the Soviet Union, when there was a famous Francis, on Francis Fukuyama.

Dr. Bao: Very well known, wrote "The End of Democracy", right?

Suji is right, to be precise, "The End of History", his original text is a question mark, "The End of History? When he wrote it, the Soviet Union had not yet disintegrated. After writing it, it disintegrated two years later, a terrible thing.

Dr. Bao: The opposite, he is actually saying that democracy is invincible, so the end of history. Democracy has become the world's only ultimate answer.

Suji right, he actually do not think this is a good thing or a bad thing, this is a relatively confusing thing. Finally this is all metaphorical, you can look at it. At that time, many scholars disagreed, saying that you are talking nonsense, how could there be one. If you think that Utopia exists, it must be an upward spiral, it can't be a sudden spiral, and then we have Utopia.

Dr. Bao: Yes, because we suddenly ended, and it was a good end.

Suji that this so-called this Utopia in theory was born, and you do not have much to do with these things, just some originally thought to be Utopia social experiment it seems to have failed, right? So that we actually Utopia on the instant, not such a thing, on a lot of controversy. But there has been no way to say where this thing is wrong or there is a lot of controversy. 00 years, accurately 99 to 01 there is a science network bubble to dot-com bubble, which is a great a crush.

Dr. Bao: Network bubbling.

Suji, regardless of whether the stock or what industry are crush off, it later indirectly led to the Fed must go to a large number of market stimulus.

Dr. Bao: Fed.

Suji right, the Fed it will go to cut interest rates, it will go to do a lot of means, and later caused a series of problems. In 08, it formed a bigger crisis. We call it the subprime crisis, is dominated by real estate.

Dr. Bao: What does "Radical Market" have to do with us?

Suji's "Radical Market" actually doesn't directly put these two things in it, but it appears numerous times in the book to talk about them. It seems that neither of these two things are working, and that none of the major systems we can see are working; it's just that we keep swinging in two directions. It's just that we keep swaying in two directions, and we suddenly say we're Utopia, and then two days later we get slapped in the face. It's very serious, like occupying Wall Street. After two days and say that place and good, that place and good, or this place and good.

Dr. Bao: The Fukuyama school also says that the end of history is long live democracy.

Suji we don't have to do anything anymore.

Dr. Bao: But there is another school of thought that says the end of democracy and the collapse of capital.

Suji, you have this left-wing philosopher Žižek in Europe, whose masterpiece is "The Sublime Object of Ideology", who went to participate in the occupation of Wall Street. That year directly said that a moment, while the birth of bitcoin he would say. (30:00)

Dr. Bao: October 31, 2018 or something like that.

Suji is right. I think it's a good idea to have a good time, but I think it's a good idea to have a good time. We are very confused, saying that we should, in the end, is not encountering any problems? Should we not, should we choose what way out? You only armed your head to be able to do the right thing in hand, otherwise you will be in the wrong direction further and further away.

Dr. Bao: Yes, armed our brains, yes.

Suji's theory of Radical Market is not necessarily right, but it is a wake-up call for us to think about whether what we have taken for granted is wrong. If it's wrong, will we go further and further down the wrong path? Or is it worthwhile to improve it a little and move forward in the right direction?

Dr. Bao: I love the concept of arming our brains. Because when you were talking, I was trying to follow up on what you were saying and challenge you to say it. I said democracy is like Starbucks coffee, it's not the best, but it's there. We always hope for a third wave revolution, like the Blue Bottle Coffee. I mean, metaphorically speaking, the author of "Radical Markets" is just unconvinced. He thinks why is there only Starbucks in the world? We should have the next wave of mainstream, or the next wave of something. I was going to challenge you and say, "What's it to me? But I think what you said to some extent has answered my question, that is, when these things are changing, you can not move, but armed with their own brains is this society in the time of change, for you to say, for all people is the greatest protection and investment.

Suji is like this.

Dr. Bao: And even this armament does not say it is the development of weapons in the head, it is a defense, you can say it is a defense.

Suji is an empower.

Dr. Bao: Empower your brain, your thinking, Radical Markets is a very special kind of thinking, because there is such a famous, so cool Chicago school heir, I first gave him a new nickname.

Suji I think the Chicago school does not necessarily like him.

Dr. Bao: But ANYWAY, the youngest professor in the Chicago Department of Economics, since he is willing to jump out and say that the Fukuyama school is also wrong and the anti-Fukuyama school is also wrong.

Suji actually yes, I have another author of him.

Dr. Bao: Posner.

SujiPosner his dad is the chancellor I remember.

Dr. Bao: Yes, the record holder for the most cited American law paper is him, his father.

Suji and their family should have been doing the law.

Dr. Bao: A family of jurists.

Suji is right. You will find that often he is very distressed about the law, saying that it's not right for me to sentence down or not to sentence down. Of course, he may have different ideas, but he will say if there should be some third way or if there is one.

Dr. Bao: The third way. Then we have several big rules or general directions in this third way. Too many chapters we do not speak one by one. We have already explained on the last time, I myself like the concept of Data as Labor. I have my own set of instructions, said in the future slide the phone will be able to live. A little like our old age pension, that is to say, your previous work, you can receive afterwards. The company's main business is to provide a wide range of products and services to the public. Can you tell the audience your version of how to explain this stuff?

Suji on the first it is sure that it is an asset.

Dr. Bao: Data should be our asset.

Suji is right. In the case of assets, we follow the theory of the left, or in fact, all economic schools of thought agree. There is a theory called the value of labor, in fact, mainly Marx inherited from Hegel's side, and then he developed it. Then there are some rightists who may not agree with it, but we all agree that some how created an asset, that I am in labor. I should be paid for my labor. As for who pays me is a different issue. Then this problem will eventually become a very simple problem, that is, the labor issue. If we see the data as an asset, then we think the user is the labor, the platform is the big capital.

Dr. Bao: Because we are scratching our phones every day, pressing buttons, shopping. even I am spending money, but I am creating data. I'm spending money while I'm creating my assets, right?

Suji right, this thing is not so obvious at the beginning, at the beginning there is an Italian scholar, we can all paste out the name, paste in there. He has mentioned a word called digital labor and play labor. play labor means that I thought I was playing, so happy, instagrm a lot of beautiful girl photos I play on that side, in fact, you are working.

Dr. Bao: Yes.

Suji right, you are working. digital labor, It means that this thing is completely digital. That they actually could not find any examples to cite at that time, (35:00) is in the ninth years to the beginning of the second millennium, because at that time the science network bubble everyone will say you this Internet is a fraud, so you do not mention this kind of thing.

Dr. Bao: We are now talking about air money, many air networks back then.

Suji right, air net, air of this dot com on the list, they actually do not have any examples to prove that this thing is right. He is a political economics researcher, he is not too many things can go to study.

Dr. Bao: It's not the same now.

Suji right, there in fact Glen his book inside have written, some relatively large turn, including Amazon since the beginning of AWS and AI, there is a thing called the Turkish, Mechanical Turk.

Dr. Bao: Mechanical Turk, very famous.

Suji that I first talk about is my work in that AI company at that time. Because we do unmanned vehicles, we actually do unmanned vehicles or do any AI are aware that the machine is very stupid. It does not know the red light and green light, it does not know the car is a small sedan, or a large car. It is a lot of different kinds of people to help us mark the data.

Dr. Bao: mark data. That is, this picture on this is the red light, that picture on this is the green light to teach the computer.

Suji right. That through my hands, that is, I know there is a lot of money to those people to mark data. At least in the early. Because it is necessary to accumulate some mods to do. There are all kinds of people who mark data, that we are still a relatively good company. If you go to see M Turk, is Amazon that platform, you will find that the above are Indian, Indian and Southeast Asian people. Then they actually English may be their second mother tongue, he is very hard to say to you, I help you mark data. And then the world's big capital is now, behind AI is a big topic, in fact, even Wall Street is involved **(36:38) are involved. This money flows into startups, and flows to them every day*** (36:42). This is the first big example, this example.

Dr. Bao: It was put in the book.

Suji right, very obvious, because the discerning eye will be able to see that this is a transnational employment of a thing.

Dr. Bao: But they have received money, that's fine, no problem.

Suji is right, but the problem is, for example, that I am a person in India and I can't do anything but drive. I'm a driver, and I'm desperately trying to mark data. I say this place is against the traffic rules, this place should stop. After marking a lot, I earned money from the Americans, so happy.

Dr. Bao: Yes, it was a pleasure.

Suji then assumes that this AI company made a model that it sold to Uber, which cooperated with Uber, and that Uber entered India with this model. uber made its own scheduling algorithm and gave Uber this kind of unmanned car, and this person lost his job, and this person lost his job by marking data. The company's main business is to provide a wide range of products and services to the public.

Dr. Bao: And I'm next door to the old king of the car, he did not earn the 50 cents, he also lost his job.

Suji is right, that is, you are not a worker thief? You're an Indian, but there's no way.

Dr. Bao: I not only harmed myself, I also harmed my own kind.

Suji Yes, that could mean I'm not driving, I'm not a driver. I'm just a convenience store employee. I can also drive, then I will bid the car. The old king lost his job after ten years. Then the old king was a driver, but he marked the tea and water, drinks and sandwiches in the convenience store, and after ten years I was unemployed.

Dr. Bao: There are also cross-category thieves.

Suji is right, so we make each other lose their jobs, so we make everyone lose their jobs.

Dr. Bao: I understand this.

Suji this time many people feel that you exaggerate, AI companies on the bubble, AI is also a bubble. ai many companies closed down, on the fraudulent money.

Dr. Bao: Exaggerated, yes.

Suji now you go to see Tesla actually opened very well. tesla is more ethical company, some companies may, I actually do not understand too well, because left this industry for a long time, but we can go to check, there are many problems. andrew Yang, is now a candidate.

Dr. Bao: 2020 U.S. Presidential Candidate.

Suji is one of the more interesting candidates in the Democratic Party, and has some support.

Dr. Bao: Not officially a candidate yet, but in his own party's primary election.

Suji, he is also considered to be a relatively supportive person in the primary election, and is the only one of Chinese descent. He has talked about the driver problem, he did not say I was so detailed.

Dr. Bao: I think it's great. You speak I think very much like that everyone understands the butterfly effect, because there is a movie everyone understands. You this is simply, this is called what effect. It is a kind of data sold, we should take a name, we understand better, get a what the polar bear effect or what.

Suji is supposed to take a name.

Dr. Bao: Because it is equal to a big circle, I never knew, I have a wing, I marked a point, the mouse marked a point. Ten years later, I lost my job, and I made 50 cents on the dot back then. This driver may still be receiving money, but there is also the car dealership next door to him, he simply did not receive money and lost his job.

Suji vs.

Dr. Bao: So in fact, Data as Labor is actually describing the problem, and its possible solutions (40:00). One of the possible solutions is dispatch. We have limited time today, so we can't seem to finish talking about it. And we want to create a special case today is to say that for the first time, there are issues in the program that Bob's friends can't finish talking about. We originally wanted to talk about the 996 ICU today.

Suji we can next time.

Dr. Bao: Yes, next time, we'll learn from Lao Gao and Mal, it says every time, and we'll create another program to talk to you after this. The last little bit of time for you to talk about Maskbook.

Suji is good.

Dr. Bao: Because it's being promoted now, right? Tell us what Maskbook is? How does it relate to Facebook?

Suji is like this, I have just talked about the relationship between Lao Wang and driver unemployment and Amazon, as well as the relationship between AI companies. This time it is still relatively obvious that there is a person who is marking data, Lao Wang or whoever. He has got a small amount of money. I just said that he may not get the money at all, you do not know they are working. What does it mean? It means driving, marking the car, it's the person, it's a task that seems very clear. I have written in my book review, how do you give a function, this function has an analytical solution, the content of this analytical solution is the relationship between the degree of Alex likes cats and the degree of Bob hates dogs, is not very complicated? It looks like a lot of nonsense, that is, what does it matter to me? Because it has nothing to do with economics, right?

Dr. Bao: Yeah, what's that?

Suji But think about it, if Alex is particularly fond of cats, there is a special function between his love of cats and Bob's dislike of dogs. Then the relationship between Alex buying cat food and Bob doing what he does, for example, supporting an anti-dog organization, is it related? Does it have money?

Dr. Bao: It is possible that I po text every day, I po the cat stuff, but because of this function, it is likely that it can also determine the possibility of Suji buy dog food.

Suji Yes, this is very exaggerated, this I actually gave a particularly strange example, but you will find that Amazon has said that the mother of this child did not know she was pregnant, he knew you were pregnant.

Dr. Bao: Just because she may have looked at the chocolate beans twice as many times as usual, touching the bag of potato chips twice.

Suji Yes, if you think of it this way, where exactly is this kind of data input? For example, I see a cat is very cute, I give it two likes, in fact, you are working. The problem is this, we are driven by a so-called social capital social capital.

Dr. Bao: So your Maskbook is actually trying to allow people to wear a Mask, to play. That is to say, I post on Facebook, or I post on Twitter or write on other social media, but he doesn't know who I am, or he doesn't know what I've written?

Suji he can know who you are, in fact, both can, both can do. But I actually think it's impossible not to socialize, not to use social network already.

Dr. Bao: That means you think people should continue to use Facebook to continue to use, it's okay. But you helped him make a barrier, right? Your Maskbook seems to allow people to write articles, friends can understand, but Facebook can not read.

Suji right, I also do not want to do a new, I also do not want to do another Facebook. I will say you installed me this thing, it can be a plug-in, it can also be a client on the phone. You can go to Maskbook.com to see, in the browser can be installed.

Dr. Bao: After the installation?

I'm not sure if I'm going to be able to understand it. If he can't understand it, I'll continue to play in a huge factory that he provides.

Dr. Bao: Play, but not become labor.

Suji is right, because now he has made a huge factory, which has more than 2 billion people in the social contract. He lied to you and said you were playing, and then you played while actually becoming labor, we told you that you continue to play? But you will not become labor when you play.

Dr. Bao: This is not just play labor, this is soical labor, we are in soical labor, but we are labor. so your Maskbook will allow us to speak among ourselves plus a layer of encryption, it will become a string of things that others can not understand, but we just have each other our friend private key, you can see.

Suji is actually a cryptographic scheme, it's not that complicated. It is a very innovative cryptography thing.

Dr. Bao: Very cool, and it sounds like there's been some unfair treatment, right?

Suji vs.

Dr. Bao: Can I speak?

Suji can speak, in fact, we can speak, because we are July 5th in Taipei to have a small launch, exactly is the United States time July 4th, Independence Day, the founding of the country, very famous holiday. It was a coincidence. After that release, suddenly everyone thought it was cool, and Vitalik had retweeted it.

Dr. Bao: Snowflake like, blockchain god Vitalik retweets, then what?

Suji and then also Tim Berners-Lee also supported us on a d web (44:51).

Dr. Bao: The man who invented the Internet.

Suji we think is very good, in fact, not many people use it, only about a thousand people (45:00). It's a drop in the bucket compared to Facebook. But these people are very like us, so they keep sending this link.

Dr. Bao: What about the results?

Suji turned out he would ban this link in mid-August.

Dr. Bao: Who?

SujiFacebook.

Dr. Bao: Facebook will put you on this secret mask generator

Suji it has no way to ban the mask, because the mask is a cryptographic thing.

Dr. Bao: But it makes you ban out of something.

Suji it banished the url, which is maskbook.com.

Dr. Bao: As long as I share maskbook.com to my friend, then he will say you passed a link that can't be passed?

Suji right, you can try, very funny, you can instagram, instagram is also, because they are a family, instagram and Facebook, and facebook messenger, except whats app, because whats app encrypted, these three inside you have to play maskbook.com or maskbook.io are our links.

Dr. Bao: And then it doesn't get out.

Suji can not send out, a box will pop up saying you can not send this link.

Dr. Bao: This is simply a penalty, and it's only an obvious penalty, and we don't know yet if Facebook will press for a hidden penalty.

Suji in fact I think it punishes that we do things more right, right?

Dr. Bao: Yes, I think this we also set up another episode to speak, there is an opportunity to speak. But this is actually true, we go back to listen to the Po Bo friends said there is a centralized words horror story, we have told a total of six, you this seventh. Very scary, I think it is the crown of horror stories. Because never thought, you engage in a new web page, service, of course he will have his reasons, he said your name is very similar to la, you are doing some encryption decryption something, punk, he must have a lot of his reasons. But we have to go back and think about it, does he really have the right to do so?

Suji he actually this thing, last time there was talk. It is not the government, it creates a commons, common space, it does not need to give any reason, this is the most terrible place, we have complaints.

Dr. Bao: Yes, it gives you a complaint button every time.

Suji is right, you complain it will not reply to you, why? Or he can refuse to reply.

Dr. Bao: It will reply with a can.

Suji is right, because it is a private company.

Dr. Bao: No one punishes it in this area, right? It's hard. It's hard to talk about it. But just for fun, Suji has also participated in some videos of TV shows on the internet, let's find out about it.

Suji can be searched, should be searched a few more times will be searched.

Dr. Bao: Yes, very interesting people, we are really, this is my fault, I can not believe that only one episode was scheduled. I believe there are several episodes to come, you will come more often afterwards, right?

Suji can, after often.

Dr. Bao: Yes, this radical market this movement in Taiwan is quite active.

Suji everyone has been very concerned.

Dr. Bao: Yes, so I hope Suji can come more often, OK, and bring us some very dense information. After that you can also put some of the fun information you just mentioned, packaged and organized. We'll put it on this explainer page at the bottom of our SoundOn channel, so you can study it properly. All right, after listening to today's episode, if you want to review the five key points of the radical market, please go to the previous page and click on the five key points of the radical market. After this episode, we will go to the next class.

Radical Markets in Taiwan:
Extended Reading with Local Perspectives

Thank you all for listening to today's episode of "What Dr. Bao's Friends Say" and we are very grateful to Suji for coming on the show today. If you like my show, remember to subscribe to my show on SoundOn. If you listen to this show on other platforms, please give me a 5-star rating, leave a comment or press the little bell to follow the subscription, and I'll see you on the air next time.

* Note: Some of the text in this chapter is marked as time code due to dictation problems, so please go to the original release channel for further information.

10. BBFT PODCAST #EP23 *"LIQUID DEMOCRACY, HOW MUCH VALUE WE HOLD?"* FT. KIN KO, LIKECOIN CHAIN FOUNDER

Originally on SoundOn & YouTube
https://youtu.be/P69AS9ORPmM

Dr. Bao: Recently, when I was watching YouTube, he would recommend me a few videos that he thought I would most likely watch. Then, I turned it off that day, and then I saw a video of Audrey Tang. It is Audrey Tang, a remote control robot to participate in a news report of the United Nations.

Gao: It's been a long time.

Dr. Bao: two years ago. As a result, I went to the bottom of the message, to see. It was six days ago, five days ago, three days ago. The first time I saw a YouTuber, I saw a lot of viewers find themselves being manipulated. I saw a lot of viewers find themselves being manipulated.

Technology, innovation, entertainment. All kinds of new and interesting in Baobab friends say. Hey, did you listen to What Dr. Bao's friends say?

Dr. Bao: Hello, welcome to PoBo friends, I am Dr. Bao. Do you think the world is fair? Let's put it another way. When we go to a restaurant, we all want to get the best service. And when the service can be quantified, it is like the rating stars on Google. Five stars plus more reviews seem to be more credible, the more likely the restaurant will attract more customers. But will this model of quantifying value for actual reward really create a virtuous model for a sustainable cycle? Will Google get the most benefit in this case? Or will the stores that are more familiar with Google get more benefits? Or will the consumers who are more familiar with the stores get more benefits? Is it possible to create a more balanced and fairer mechanism for this world? We are joined today by my idols, my idols. I'd like to say hello to Kin Ko, the founder of the Appreciative Citizens Foundation and Appreciative Coin.

Gao: Hello, everyone. Rebecca Ko, from Hong Kong. How many idols do you have? How many do you have?

Dr. Bao: My idol on you and Tom la, OK. Yes, we do not have many idols.

Gao: It so happens that my idols are also you two.

Dr. Bao: from Hong Kong on the two of you. No, of course there are many more, don't offend others. I don't know if the audience will find our voices particularly magnetic today? If so, it's because we are recording today at a rather special time. It's at a time when we want the world to be healthy and better. If you want to know what the situation is, it's that we're all wearing masks. Anyway, Kin, would you like to tell us a little bit about yourself? I'll do so later. But I'd like to say, do you have your own version of the story? You usually keep a low profile for too long.

High: I am in the so-called official statement, because often need to register many accounts, may need to write some very short introduction. I am an earthling. Then, I will be very concerned about a principle is that I hope the humanities-based, technology for the use. This is my very short self-introduction.

Dr. Bao: We usually dig into the details of our guests. He said he graduated from the Chinese University of Hong Kong in 1997 with a bachelor's degree in computer engineering. And then later took a master is information technology management. So you say, humanities-based, science-based, right?

Gao: Yes.

Dr. Bao: But your background is more engineering oriented. Where did you get this humanities-based approach from?

Gao: In fact, you introduced me this way, I myself are not too good to say that I am studying engineering, because my technology is relatively bad. And then it's been a long time since I wrote code. And in college, in fact, almost want to switch to that social science. So that background is actually quite mixed. On the engineering is the computer, and then politics and is sociology.

Dr. Bao: In fact, I also want to tell the audience that Kin is the most inside my friends.

High: The old one.

He told me that he usually walks a lot, and then he doesn't bring mobile power. This seems to be a bit.

High: about low key what relationship.

Dr. Bao: In short, he is a very low key. but he has a very sensational past. We are now going to talk to you. From 1999 to 2017, you did an amazing thing. Do you want to talk about it yourself?

Gao: No, no, no.

Dr. Bao: You see he is like this. Every time I ask him to speak, he did not have no.

High: I just opened a company with a friend to do cell phone games. Then the middle is very, very long time to stay in the mainland (05:00), there is no special what. How did you describe it just now?

Dr. Bao: the astounding achievements of

Gao: No, no, no.

Dr. Bao: I'll have to tell you about it. The game company founded at that time is the hand game. Called LaQuan game, you can go to check. Should be that pull, pull the alarm pull. Broad is a door, then a live. Right, Lakoo, pretend wide. The company is invested by Tencent and Sequoia Capital. In the field of innovation and entrepreneurship, people do not necessarily kneel down when they hear about Tencent because of their own subjective position. But when they hear Sequoia Capital, they usually fall to their knees. Sequoia is one of the most famous venture capital firms in Silicon Valley. You then took which round to get their investment.

Gao: First Tencent and then Redshirt.

Dr. Bao: We can see that in 1999 to do cell phone games. Don't forget, the iPhone was not yet born in 1999. So what kind of cell phone games did you make at that time?

High: The ones in Nokia, feature phones. Now look back before you call those feature phones. Anyway, it is the hottest feature phone of the era.

Dr. Bao: In Taiwan, our feature phone is called a smart phone. It's not a smart phone, a smart phone.

High: We call it a stupid phone.

Dr. Bao: In fact, you can run some JAVA software.

High: Not yet in 1999.

Dr. Bao: So what are you going to use to write the game in 1999?

Gao: In 1999, when the cell phone and communications industry began to develop a new standard. That is called WAP. or say, there is another saying called MMM, the WWW reversed to call MMM.

Dr. Bao: I have not experienced this era.

High: Once called, there is a relatively short period of time, that is, the cell phone Internet access, and computer dial-up Internet concept is similar. If the cell phone to go online, basically you will not be able to talk. And you are not using the package to count, you are using the length of the communication to count. If you have a minute of communication, you will be charged a minute of money.

Dr. Bao: Not counting packets, not counting capacity size.

High: GPRS has not yet been launched. At that time there was no package.

Dr. Bao: that time only WAP, you just said, I do not seem to have experienced that.

Gao: very specific, I should say, when our company was founded, WAP that standard in the noise. Then there was no first cell phone. In 1999, if I remember correctly, in December 1999 there was the first cell phone is Nokia's 7110. you can fact check it.

Dr. Bao: I tell you, I went from retarded to wise, that is, from Nokia's 6600. you this?

KO: Of course I know.

Dr. Bao: It's the chubby, Symbiany operating system.

KO: I know.

Dr. Bao: You are really like that, almost the company of the zero wave of cell phone games.

Gao: Yes. But I did not say very sensational, because simply put, it is too early to do.

Dr. Bao: Well, let me give you a quick turnaround. That is, in 2017 you jump into the world of blockchain.

High: Sort of like this, it does not feel like this. But, anyway, the objective effect is this.

Dr. Bao: Okay, that's good. I'm going to give you a background. In 2017, there were a bunch of companies that jumped out and issued coins. At that time, bitcoin was at $19,000, give or take. In 2017, or early 2018.

High: 18 years.

Dr. Bao: 2017 was very prosperous. Then many companies came out to issue coins and make a lot of money. From dog coins to fart coins (08:40), and then to messy coins. Then there were a lot of platforms, and they really made a lot of good products that are still operating. At that time, there were many teams in Asia, and they wanted to make a bitcoin or an ethereum. The first time I saw this, I saw that the earliest users of Bitcoin and Ethereum were really the most popular. The first thing I saw was that the earliest really serious in doing, and issued his coin, and on doing community, there is in doing products, is LikeCoin this LikeCoin is Facebook press like that, coin is coin. coin. is not a bit early ah? What is your own assessment?

Gao: It should not be too early. However, my own, my own review in the past is generally speaking, I will do too early. However, if you simply say that the financing point of time is not too early. Maybe it is just right, a little too late.

Dr. Bao: This is actually really mysterious.

Gao: We review every year, I said the team will review every year. Every other period of time are reviewed, a year will have a big review. We look back, in fact, we feel that this way of saying that there is, that is a little bit ironic is (10:00) we feel token serve (10:03) we do too seriously.

Dr. Bao: token serve is the end of selling coins. That is, you have some kind of degree also considered to have run through ICO that section, although now ICO now sounds very like a dirty word, but you do have on the Internet to sell coins.

High: Our review is that we do too serious, and then dragged a very long. The white paper has been written for half a year. And then from the time the price of the currency is very high, has been to do the price of the currency is low may be seventy percent of the same.

Dr. Bao: So what you're saying is that if you hadn't been so serious about writing the white paper back then, and had gone online to raise money earlier, maybe the amount would have been more.

Gao: Definitely, it should be definitely.

Dr. Bao: How much money did the ICO raise at that time? How many

Ether coins?

High: seven thousand, like seven thousand two hundred more. Anyway, more than seven thousand, right.

Dr. Bao: I saw Akane's eyes widen.

High: Almost no ICO is seven thousand.

Dr. Bao: At that time, 7,000 etheric dollars. An ethereal dollar is 500 dollars, right?

Gao: We should do it, we dragged it out for a long time. The white paper alone has been written for six months. It should be from three hundred to twelve thousand have.

Dr. Bao: So you'll average five hundred. You can do the math yourself, 15,000 Taiwan dollars times 7,000. It's a little hard to calculate, isn't it? I guess it's about 100 million Taiwan dollars, if you average it out. All in all, you must say that LikeCoin is too late for you. But in Asia, it's actually in the middle? It can't be considered early.

Gao: I'm just saying that it's late for financing, but it's not very late to do this.

Dr. Bao: Tell us what LikeCoin is?

Gao: The LikeCoin question is actually quite difficult to answer. But before I say what he is, let me say what he is not. Because just now, I heard you mention that we have some kind of platform. I care about that because whenever someone says LikeCoin is a platform, I care about that. I always make it clear that we're not a platform, we're a floor. We are an ecology. You can think of us as a European Union of independent media. The European Union is each country, he will have its own sovereignty. When each country has its own sovereignty. But at the same time they can have a common set of currency. LikeCoin is probably such a mechanism. Specifically, from the technical point of view of the embodiment is that we are a chain, LikeCoin chain. then there is a token, is LikeCoin. there are also a variety of plug-in API and SDK. let different platforms flat. So the platform is our partner, not us. For example, you operate a website you can, for example, SoundOn can add a small plug-in, will enter the ecology inside.

Dr. Bao: What Kin just said is the description of version 3.0. I'll help you rewind to the LikeCoin 1.0 version.

Gao: I'm not very good at it.

Radical Markets in Taiwan:
Extended Reading with Local Perspectives

Dr. Bao: My feeling, I think, if I introduce LikeCoin to others, that is, my students introduce LikeCoin, I will use a relatively introductory way to explain. I would say, "Have you ever used Facebook?" Facebook you see people po article, you like it, you click a like. I would say, "Have you ever used Facebook? LikeCoin is to use blockchain to bring social capital and social money to the surface. As long as the small plug-in of LikeCoin is incorporated, you can click the Like button. But this like, each like that author will receive a real like coin, called likeCoin, and then this coin can be taken to the blockchain exchange, in exchange for real money. Do you feel very excited ah. Like Dr. Bao sometimes write articles, many people click like, I actually just eat dinner at night a little happier. I did not manage to eat better. But with LikeCoin, I'm telling you, Kin has told me a vision before. This is your 1.0 vision. Now your vision has become very complicated. That's the 1.0 vision.

Gao: Creativity can be eaten as a meal.

Dr. Bao: Yes, creativity can be a meal. He can earn kudos. You can go to the bakery that supports the appreciation coins and buy bread. At this time, you can get the art of creation, love and bread.

Gao: I think you reminded very good. Many people will remind me that many times, it may be too deep. And then you are better at telling stories.

Dr. Bao: That is to say, in 17, 18 years when the praise coin appeared, then mainly so that people can be placed in (10:00) Medium, or WordPress, or Blogger or some creator's platform.

Gao: Let me add. We are doing oice first, then oice. oice itself is a platform for visual novels. This is indeed a platform. In the process of doing it, we want to achieve an effect is because, ooice this platform allows creators to use a lot of material to do their own stories. If this author's story can be sold, we hope that money can be distributed to many different people who provide material. Whoever provided me with a soundtrack. Who provided me with a background. Who provided me with an item and so on. We take it very seriously and break it down. It was one dollar that I might give to thirty people. Then it probably didn't work out. Then we wanted to do it through blockchain. Because this project is too big, then instead of the focus on LikeCoin, there is no time to engage in ode.

Dr. Bao: In fact, when it comes to ode, it will talk about the trajectory of the material. This creator, you created a picture. If it is applied in the second degree, the second creation, the third creation, the fourth creation. It is possible that through the blockchain, you can share the profits again and again, or get your original creation again, when it is created twice, three times, four times, eight times, you may still have some meager income. The company has been doing a lot of things in 2019. I say this you should not understate it admit it. Because LikeCoin has gone up one hundred and fifty percent. He must be very angry. I'm talking about the money now.

Gao: No, I did not pay attention to this data.

Dr. Bao: But behind the increase, they actually made a real value. LikeCoin did an amazing thing in 2019 when it officially broke through with the total number of user members and articles.

Gao: 160,000, I think the total number of articles that have been liked is 160,000.

Dr. Bao: It's the article that has the button to collect the coins for praise.

High: Articles that have been liked.

Dr. Bao: Some of them are installed, but they are not liked. I know you have told me that there may be 450,000 installed. I don't know if they have bought bread, but they have received 160,000 coins, a lot. What about the creator?

High: Five thousand creative authors.

Dr. Bao: five thousand. Fuze to the creator.

Gao: There are also five thousand authors who have received praise.

Dr. Bao: I heard that because of Dr. Bao's sharing, the users in Taiwan. I talk nonsense.

High: A little help.

Dr. Bao: said Taiwan users actually accounted for a lot of components inside.

Gao: Taiwan has the most, a little more than Hong Kong.

Dr. Bao: It's really great. And in 2019, they did two amazing things that I can remember. The first one is that Originally Applause is a blockchain that relies on ethereum. It's a coin under the ethereum ecosystem. So is Dr. Bao, although no one knows what Dr. Bao is now.

KO: I know.

Dr. Bao: Do you have one?

Gao: I did not.

Dr. Bao: I am now on Uniswap and can go buy two of them. This is not the point. likeCoin was originally an ERC-20 coin under the ethereum ecosystem. But in 2019, they set up a chain of their own. It's also a miner of its own. It's not a miner, or it has its own bookkeeper. And their own ecosystem. He doesn't need to rely on the health of the ethereum, or death or not. It can sustainably rely on the co-managers of the ledger and the supporters of the ecosystem. It's a way to make the Kudos and the Kudos Foundation and the Kudos Republic more sustainable. Can you explain the process from LikeCoin to Likechain?

Dr. Bao: First of all, I must say that I actually like ethereum, and I have liked it so far. However, there are some application scenarios that we need, but he can't satisfy them. We have launched LikeCoin chain, which can be interpreted as we are independent from the ethereum and established a country. This is probably the concept. Specifically, why can't we meet it? For example, there are some scenarios where ethereum is particularly suitable, such as DeFi (19;34). You may have more money, you won't be very frequent, you won't be doing regular (19:42), you may be doing ** once a day (19:43). So you wouldn't mind saying that you need to have a small handling fee. You don't mind that there's a threshold, you need to have a little bit of ether to pay that miner's Gas Fee, but our scenario is exactly the opposite. (20:00) We have a lot of very, very frequent transactions. But each one is very small. Because every single like is a transaction. For a user, he may read the article all day long and occasionally click on it, and there may be dozens of transactions. The thousands of users are tens of thousands, and I should say that now it is more than 45,000 users will have a lot of transactions. It's likely that if they stay in the ether, **the Gas Fee will be higher than the fee given to the creator.**

Dr. Bao: You just said, in fact, five thousand is the creator. In fact, the one button that really goes to press the like button. The number of people who gave a like is actually super high.

Gao: About ten times.

Dr. Bao: So the daily click-through traffic is actually very high. Ethereum may not always be suitable.

High: Very unsuitable.

Dr. Bao: So you decided to go away.

High: This is the first very main reason. In addition, there may not be so well understood is that we want to operate an own autonomy. That is, with its own sovereignty. My analogy just now is also that we are independent from the ether to establish a country. We want to be the so-called on chain govern, the chain control. I hope that the people who hold LikeCoin can also participate in deciding how this LikeCoin will go on. Rather than by a small team of us. Or rather, the general development of people to decide so. So, we have to go out independently and have our own ecology to achieve this thing.

Dr. Bao: This is very cool, why? Because it's usually not easy to build your own chain. The previous, LikeCoin, Litecoin is a reference to the chain of Bitcoin. And then fork out, is forked out. It should be said that the ether is also extended out to give birth to many similar than the ether, such as TRON, such as the disappeared ** (21:56), or other High:** (21:57).

Dr. Bao: Yeah, yeah, LikeCoin didn't start from scratch. They used a cool, sort of system? This is the first time I've heard of it from you. He's kind of like the mother of all chains, the mother of all chains.

Gao: But we were really developing before Cosmos. The test net (22:26) was also launched. The test net (22:26) was also launched, and Cosmos was officially launched in March of last year. The first thing is that Cosmos is an SDK that can help developers to build their own chains with relative ease. We can consider Bitcoin, Ethereum, and Cosmos as three generations of blockchain. The first generation is Bitcoin in 2009, and at that time there was no division between the chain and the coin. Because Bitcoin is that chain. The only thing that the chain does is Bitcoin, and the only thing that it does is to make a money transaction. This is the first generation of Bitcoin blockchain. The second generation will be five years later. In 2014, there was a white paper of the ether. To me, my understanding is that the most obvious thing is that he, Vitalik saw that the chain, in addition to dealing with money transactions, can also do other transactions. And then for a technician to be able to trade on behalf of being able to remember the value of that probit and so on, it means you can do the program. Then the ethereum will allow other people to issue their coins based on the ethereum. This is a token for a particular application. then after five years, just after five years, is Cosmos.

Dr. Bao: 2009.

The first generation of LikeCoin is also based on this. The first generation of LikeCoin is also based on this to do. The first generation of LikeCoin is also based on this, and now Cosmos can let us do is application specific chain, so even the infrastructure is your own to build, he let us develop the threshold lowered a lot. Of course, the rest of the threshold is still very high, that is, someone must be willing to help you do the bookkeeping and so on, need to establish an ecology to do. The community has been built through the use of a specific token (24:33) LikeCoin, the old version of LikeCoin. That community today has about 700 sites, more than 50,000 Likers. We call our users Likers, and about 10% of them, more than 4,000 and 5,000, are creators. The whole ecology of people to support a new, a new government, a republic, this is what we mentioned earlier (25:00), we call Like land, Republic of Like land to praise the civil republic.

Dr. Bao: Do I need to talk about the translation of this?

Gao: Yes.

Dr. Bao: How do you know. I'm going to talk about the cow version of this. That is, Bitcoin in 2009 allowed people to make something like money and to be trusted. And with ethereum, everyone can make something like bitcoin, and 3.0 is the so-called Cosmos, which is even cooler. That is to say, you can not only create your own coin, you can also create your own chain to support your own coin. After the talk, so there is not relatively simple, applause to encourage. Actually, it's not bad at all. At that time, Kin actually told me that I think there is a place that is particularly interesting, the Cosmos ecosystem has its own main currency, called ATOM, although it seems that not many people are using it. I can't figure it out, you wait a moment to explain it. And then, you use the SDK to create your own chain, your own chain, but also can have their own coins. So, we all think Cosmos is a bit like a mothership. This mothership has its own circulation currency, called ATOM, and you use the resources of this mothership to create a chain of your own. Under the Like chain republic, your circulating currency is called LikeCoin, please correct me.

Gao: No, no correction. Let me change my statement. You are not wrong, but I will change my words. First of all, Cosmos can be subdivided a little, from the technical point of view he is a SDK. a little like Linux, you take it to develop your own things, your own brand, your own a deployment, you do not have to its services, is to use that set of code. The company's main goal is to provide a complete solution to the problem of the problem. The first thing you need to do is to use the token. So what does he use it for? The simplest way to understand it, we can, because what he wants to do is a so-called Internet of blockchain (27:18), now our world is a lot of websites connected together. He imagines a future with many, many blockchains that communicate with each other. That a chain with another chain to communicate, that is, LikeCoin to be from LikeCoin chain, issued to another chain in the middle of that Gas Fee, that is, you can use

ATOM to pay.

Dr. Bao: Now Like chain can communicate with other chains?

Gao: Not yet.

Dr. Bao: I ask a muggle question, if the Cosmos SDK is upgraded, do your engineers have to do something accordingly, or have you completely disconnected?

High: We look at the situation, we can. It actually depends on what features they have in that upgrade. Then we are trying to pull this feature into the LikeCoin chain.

Dr. Bao: So basically, you can be completely free of him?

Gao: Yes.

Dr. Bao: What we just talked about was actually the first thing that was amazing in 2019. We are now quickly moving on to the second event. The second thing is that in 2019, LikeCoin officially cooperated with Matters.news, one of the most popular online article platforms in Hong Kong, Macau and Taiwan, and in the Greater China region. Would you like to tell us about it? I was very excited. I thought it was a strong partnership, which I have rarely seen in the last decade. Can you tell us about the importance of this partnership?

Gao: This is good, first from our white paper back to the time of our white paper. Because, we actually want to do, at the beginning I introduced LikeCoin said I am not a platform, we are doing infrastructure. However, before the popularity of people do not understand, but also do not necessarily use your infrastructure. So as a last resort, you need to do something to tell people what can be done on top of this infrastructure? So the white paper, we have written to us to build an application, a Dapp, blog chain. blogs.

Dr. Bao: I have not heard of this.

Gao: The white paper has. But we actually do not want to do. Do not want to do is not to say that the technology, but that I believe in myself, especially respect, in fact, editorial operation is a lot of professional in the inside, I think this is not suitable for us to do. Moreover, we do infrastructure, if at the same time to operate a platform, is not something we want to do. Because with your 1.0 method to say, LikeCoin is what? (30:00). We're not saying we want to be a platform. And then if you want to earn coins, you come to our platform to post articles, or rather post pictures. That's what a lot of businesses do. But we think we shouldn't do that. So it is very stressed, always very stressed that we are doing the bottom. Then ask the other you do platform intervention it. That Matters is one of them. I just said there are more than 700 of one of them. But it is a very special one. Because his degree of integration is the deepest. Specifically, each Matters user will have a like id and wallet. Each article will automatically have a like coin button, so that everyone can clap, and so on and so forth. Many other sites are wordpress will also have that clap button for each article. But not every reader will be able to clap. Because the reader will not necessarily have like id.

Dr. Bao: So the way it works now is that anyone doesn't need to know blockchain, just see the article, there's the like button which is the like coin button, just give it a click and that's it. The author will receive like coin. like coin is now really convertible to money on blockchain exchanges. If I remember correctly, the average price in 2020 now is 0.15 NT for a like coin. I don't know if that's changed in the last few days.

High: Almost all in the zero point, 15 to 20 cents almost.

Dr. Bao: Then I remember correctly, usually the article is not too badly written, will be at least an article almost more than thirty likes. If you post it in the right place. So after writing an article, I think it's not a problem to earn five Taiwan dollars. And if there is a long tail effect. We think there's a very good chance that one article will be worth one loaf of bread. Right, founder?

Gao: You may not have a problem. However, to be honest, we do not want to be, should say that the state of the world now is only the most popular of those who can earn money through creative work. I can not promise, I dare not go to promise is to earn a lot. But we hope that there will be. This is very important. From zero to zero.1. The meaning is different from zero point one to one.

Dr. Bao: One of the benefits of the blockchain world is that I can give you a dollar, or even zero point one dollar. But in modern finance, using a bank passbook. In fact, this kind of micro-payment, or micro-appreciation or sponsorship, is relatively very difficult. So, in fact, another question that may be raised is, where do the so-called LikeCoin coins come from that the creator receives and is praised? Because when you see a good article, you see a clap button with a token of appreciation. You click it five times, and you don't have to pay for it. But the creator receives the money. How is this magic done?

Gao: I think we can explain it from two levels, one is the operational level and one is more conceptual level. I will talk about the operation first. We just said, I just said to our users I call him liker, which is itself is free. But he can choose to become a Super Like (33:26) what we call an appreciative citizen. He needs to pay a hundred and fifty dollars a month to become a rewarding citizen. This 150 NT is not for us as a team, but for all the creators. So after I give a hundred and fifty, for example, I like different things that month. For example, if I like a hundred and fifty articles, the author of each article will receive one dollar, which corresponds to one Taiwan dollar in appreciation. If he likes 15,000 posts, each one will get one Taiwan dollar. He can take as many as he wants, and he can take as many as he wants. We will distribute them evenly for him. In addition, we have a matching fund. Whenever a citizen gives a hundred and fifty dollars in appreciation. We will match the fund with a hundred and fifty percent of the LikeCoin, which is one hundred and fifty Taiwan dollars. The 150 paid by the appreciative citizen is completely governed by his own praise, and no one else will influence it. However, the matching one is influenced by everyone, and everyone is influenced by the likes. So we can imagine this fund as an existing government, for example, the Taipei City government may have some funds to support creative works, and he will have some teams to judge different applications and whether to grant him some money. We also have this kind of fund, but we put the power completely down to all likers, and each of his likes is equal to a vote. We count the votes once a day. (35:00) and then send that fund out once like this.

Gao: I'll translate it because some of the terms may be, as you may know, some of the subsidies from the Taipei City Government for entrepreneurship, or some of the subsidies from the Ministry of Economic Affairs for new ventures. But there are benefits to these grants, but there may be some status quo, that is, they are decided by some members. It's not a bad thing, but in the world of the Republic of Applause it's up to the reader to press the clap button with his index finger. After one day or one, after one time, anyway. This Applause Republic platform will know which article, like the best new company or entrepreneur, will be distributed equally from this fund. He gets more pats on the back, and he gets more praise money. And this is the best part for me, in fact, the first time I know about this is that I know for the first time that if a person pays $150, that's $5 per gold to become a praising citizen. It's a paid appreciation citizen, but it's a VIP citizen.

High: We call it taxable. We put metaphorically put as taxable.

Dr. Bao: Taxes, not Nazis, taxes, we have masks today. We have masks on today. We pay a tax of one hundred and fifty dollars a month for the appreciation of citizens. That's why the appreciation money is called the Appreciation Foundation. What do you call it, the Appreciative Citizens Foundation, and this foundation will also allocate another 150 NT dollars at the same time. That's really great.

High: This money is the front, you just mentioned that ICO has a part is used to support this.

Dr. Bao: So really, everyone is welcome to register as a tax-appreciative citizen.

Gao: There will be some other benefits as well.

Dr. Bao: For example, you use Dr. Bao's referral link, Dr. Bao is going to get, as if you will get a little benefit, right?

Gao: I will have some benefits. For example, we have just launched will have discounts in some bookstores to buy books.

Dr. Bao: No. What I care about now is that people use my referral links.

High: Yes, there will be.

Dr. Bao: I will have the benefit, he will have? Because Ming Zhen feels like he's going to pay for it later. I am now rushing to give him that link.

Gao: You will have that, there should be the equivalent of five dollars.

Dr. Bao: Kudos coins. Wow, it's super profitable. So Ming-jin, sign up with my link, get your referral link, and then ask your boss to sign up. This way you can also earn five dollars. And in fact, the point is that you can certainly save this five dollars, you can take to vote, you can also take to sell. The most important thing is that you can actually take it and give it to other people. Next, they'll launch you with a really great article, and you can use the five dollars you receive in your own wallet, and you can give an additional five dollars to an author for more.

High: Like pay.

Dr. Bao: LikeCoin with like pay of this mechanism. So we are going to talk about December 2nd, 2019, there is a third, we just counted one, amazing, I like a lot of things, called Liker Land app officially launched to start operation. Please take out your cell phones, whether it is iPhone, android, or even a browser, you can browse it. Go to Liker.land.

High: The URL is this. If it is apple store and android should be liker land or search likecoin can also find.

Radical Markets in Taiwan:
Extended Reading with Local Perspectives

Dr. Bao: It's a praiser, a republic of praising citizens, the concept of an island. It's an app, it's an app that I think is the easiest to use wallet in the world today in the blockchain world. You don't need to understand it, you just need to have a Google account, or you just need to have e-mail and you can sign up. There's nothing in there that you can't read, no mnemonics, no mysterious private keys, nothing at all. After you sign up, you become a republic of appreciative citizens. You go to write articles, you go to the click, there will be your record on the above very amazing, how to do it can you explain to everyone?

Gao: Yes. I'm glad you said that, because that's really one of our core designs is that we want to give creators the benefit of not having to know blockchain at all. So we're trying to put all those technologies on the ground floor, so if the audience really takes out their phones and looks at the app. From his introduction to the download, (40:00) to the registration account and so on. You won't, you won't see the word blockchain at all, and then you won't see crypto, you won't see exchanges, you won't see wallets, you won't even see the word LikeCoin. We positioned it as a reader, so if you want to read the news or read the story, you download the app and read it. But all the details are not gone, it's right underneath the bottom. He is a bit like our now popular MacOS, in the tenth generation when he changed the bottom layer to Unix, but the surface is completely invisible. We are probably designed with this in mind. Those who can take away, should not force ordinary users to learn things are hidden underneath, to help him deal with. And then just use a very ordinary packaging. In fact, if you are interested, you can join our meeting every Monday morning. The work plan is very transparent, and then you will know exactly what we are planning for the whole year.

Dr. Bao: The reason why I really want to talk about this is because I encourage all my friends and listeners to talk to their friends and friends of friends. I really hope that we can find a place to absorb information other than Facebook. I myself really feel very afraid of this kind of centralized information platform recently. Because Facebook's algorithm determines what all people are watching these days. Let me tell you a funny one. I was watching YouTube recently, and after I watched a video to turn it off XX. in the YouTube app, it would recommend me a few videos that it thought I might watch. Then, I turned it off that day and then saw a video of Audrey Tang. It was a news report of Audrey Tang using a remote controlled robot to participate in a United Nations conference.

KO: I've seen it. It's been a long time.

Radical Markets in Taiwan:
Extended Reading with Local Perspectives

Dr. Bao: Two years ago, and then I clicked in at that time, and then I looked. This is not an old article? I went to the bottom of the comments to see, you know what the people in the comments are writing? Six days ago, five days ago, three days ago, everyone wrote that YouTube's algorithm was amazing and that everyone was brought here by the algorithm, right? YouTube, for some reason, has recently started pushing Audrey Tang's content. In fact, I know the reason. It's because the Boone Nightly Show has interviewed Audrey Tang. And she is not a person that young people and YouTuber channel consumers would have known about. So YouTube detected this incident and suddenly discovered that Audrey Tang is a popular content. The company's main goal is to provide a platform for the public to learn more about the company's products and services. This thing is really super scary.

High: Also good Audrey Tang is very nice.

Dr. Bao: And it's good that I see a lot of viewers at the bottom who find themselves manipulated. They never thought that YouTube would recommend a two-year-old video for them to watch. They also never thought it was important and interesting to see Audrey Tang use the robot to go to the United Nations. But they saw it, because it was the first one on the first page of the home page. And then that video, now a million views. A lot of people at the bottom said what is this? But many people know that this is the algorithm. So this thing is super scary, in the Liker Land app, you can see the good content without this kind of operation. You can choose what you want to read, and you can choose how much you want to support such authors. In the next season, the third season. We can even decide to have a separate section by your own track of likes. You can read more content from people you like. And when you want to get out of that bubble of preference, you can go back to the river where you can show all the articles. It's very exciting to see, and Kin wants to leak a little more, 2020. After three amazing events in 2019, LikeCoin and Appreciate Republic have some plans for what's next.

Gao: I can say that what we are going to launch in 2020 has been planned for a long time. One important thing about us is that many people have the misconception that we are a Hong Kong-based or Chinese-speaking project. But that's never been the case. Of course I am very fond of Chinese, but also very fond of RTHK. We are a project that has no borders, it is for creators and people who like content all over the world. It is simply that we use Chinese first to implement some functions in Taiwan and Hong Kong first. Because the project to do the Internet is to do a good job of that so-called proto market * (44:55). First of all, those who should fix the bedbugs (45:00), debug good. The rate, conversion rate, and so on are well tuned. And then this year, perhaps the most important thing for us, in addition to those features that have been planned to launch. It is to bring the whole LikeCoin project to other languages. Including Japan, Korea, English and so on and so on different cultural places. But this will not, not by us. I said, we a small team to do. Because originally we are the open source. And then also many volunteers to help us do this thing. What is important to us is to organize our APIs and documents and so on. And then, I think there are still a lot of people who are interested in bringing this to different platforms around the world.

Dr. Bao: So internationalization will be your next step, and you will step out of Asia, or Greater China, or the Chinese region.

High: I think he is a natural thing that will happen. We have to do a good job or, or that we put the infrastructure in place on the good. We may have done many things before because we went too fast, and then may not have organized well. Including documentation, etc., and those APIs actually exist, but there is not enough documentation for others to use. So on the LikeCoin chain you just said launched, and then we did a demo ourselves, that matters can be so to integrate into. The latest in the busy to put together those APIs, on who can make his own matters. may be the Japanese language matters, Korean matters, English matters. open Medium, can also be a special not for the text, can be for illustrations, illustrations, can be for photos and so on different communities. It also uses the same basis, that is, the praise. It's actually two. We will look at it as two. One is appreciation. The other is one I would call Like pay.

Dr. Bao: Like pay.

High: is a direct transfer of funds

Dr. Bao: It's very exciting. As I was listening, I was thinking about a philosophical novel called Atlas Shrugged, a three-volume masterpiece. It depicts a situation called what would happen if the thinkers of the world were to go on strike. While I was talking, I was imagining in my mind what would happen if the creators of the world, creater, went on strike. Because they don't get enough money, they are accidentally yellow-labeled, they are accidentally blocked, accidentally taken down, what kind of world would it be? I feel like Kin is doing something similar to LikeCoin's praise of the Republic, which makes me feel like it's of the creater, by the creater, for the creater. Speaking of which, can you tell us a little bit about the reason you created this term, called Republic. Is there a deeper idea behind it?

Gao: Deep, huh?

Dr. Bao: I also feel very virtuous to have coined this adjective myself.

Gao: I understand it this way, because someone has created something of value, which is content. So we can generate more, or cast some LikeCoin to multiply this value. Then we generally few people will understand this way. Because we have a concept that only the national government has the power to mint money. So we generally find it very strange. Even though we agree that there is an article written out. However, we do not feel that we can mint some new currency to carry the value of this article. That's why we praise the civil republic, but from another perspective, we are actually doing this. Therefore, in my world view, I would regard the traditional state, that is, the nation state, as one level. Another level is that everyone lives in different countries in different cyberspace at the same time. This one of them we are interpreting is to praise the civil republic. There is no conflict between these two, it is my body, of course I live in a certain country. The physical things in my life, such as where I live, transportation, and now I have to wear a mask, are managed by the governments of these nation states. However, there are many things in spiritual life that cross national boundaries. Like Facebook and Google, which you mentioned earlier, all the online stuff is cross-border. So these online communities, (50:00) which may have different characteristics, may be managed by some countries in cyberspace. We will measure how much value a piece of content has generated and generate more LikeCoin to distribute to different creators and other related people.

Dr. Bao: In fact, just now, when Kin was speaking, I had a picture in my mind. It is that we are talking about the future is likely to be a multi-nation, or talking about called multi layer of nations or nation shopping. is that we are now feet on the land, I am a citizen of this country, I do not have a second choice. Unless you use what investment immigration or other methods. In the future, there may be another layer called digital world, virtual world. in this world, you may also be able to join another mechanism, ecosystem, ecosphere, become a citizen of this ecosphere. The picture in my mind is that, in fact, in the foreseeable fifty years, human beings will become an interplanetary species. In other words, before we die of old age, we will see the first generation of human beings, who were born with no land under their feet. That is, he will probably be born in space, he will probably be born on the moon, he will probably be born on Mars.

It is very likely that his nationality of origin is a citizen of several worlds. This is possible. Before the countries on Mars are determined, and before the borders on the Moon have been drawn. I think that in our lifetime we will see the birth of digital native citizens without a physical nationality. Until then, we have to build a country for this child that belongs to the digital world of his choice. This is actually what I find very interesting. So, let's go back to a more concrete point, that is, the validation of people that we just talked about. The republic of praise that we just talked about is a bit like this Caesar who built Rome, but with a Senate. But Caesar chose to abdicate. You are the one who started the territory but voluntarily stepped down and let the Senate decide the policy. The Senate is currently said to have seven organizations.

Gao: Yes.

Dr. Bao: It's like Kojima (52:38) talking about I am the swarm, he's talking about that Starcraft. It means that there are seven senators in the Senate of the Republic of Praise. But this patriarch is not a person, he is an organization. The current verification of the seven verifiers, that is, the seven patriarchs, which are the seven organizations? I know there is the Blockchain Research Society of the North University of Science and Technology. Who else is there?

Gao: Yes. Basically, they are all RTHK media-related, or public interest-related organizations. Especially open related, including the North Science University you just mentioned. And then we may have seen, or at least heard the Hong Kong position of the news, independent media, the news, and COTA, COTA is an organization that does open culture and technology. There are also several purely technical ones, such as FORBOLE, a blockchain company specializing in Cosmos technology, and so on.

Dr. Bao: So we actually have these patriarchs to help us hold on to this very promising and ambitious republic that the abdicated Kaiser has created. And I don't know if I can report, that is, afterwards, if someone also wants to be a witness, is it possible now?

Gao: Of course you can break the news, because it was already on the Internet, but not many people cared about it. We happen to be working on the first motion now. The first motion is to lower the threshold for being a witness, down to zero. This is no need to get anyone's approval, he can set up a server, the server can apply to be a witness. And then the so-called application, in fact, is not a very accurate statement, because it is not I approved or not. It's about whether he can get enough Likers to give him the pledged LikeCoin, because we'll go from seven to ten. But it is likely that there are more than ten interested units. Then only ten can get, that is, the ten with the most votes to become a valid verifier. So that is very similar to an election. We just mentioned mobile democracy (55:00), but we haven't explained it in particular. Now I will explain a little bit, it is a manifestation of democracy. It's similar to the representative politics we are familiar with. But there are some improvements is, first I have the LikeCoin I can vote, but I can elect my legislators not only to elect one, I can choose more than a few, and I can choose that degree. I'm a bit of a supporter of whoever gives him a little bit of LikeCoin this is the first. The other point is also very important, is that he does not have a, there is not a say a few years to choose once. Rather, any time. Specifically, it should be five seconds. Because LikeCoin is five seconds a block, every five seconds you can choose to put your LikeCoin from a Validator, pledged to another. So if anyone does a bad job, then those Likers will immediately run away.

Dr. Bao: I think everyone listened, may not have operated still not very able to understand. However, if you have to download the Liker Land app, you will see that there is a deposit function, a sending function, and a function called entrustment. You can entrust your coins, LikeCoin, to the unit you trust. For example, if you think Matters' position is very consistent with yours, you can entrust your coins to him, and he can help you to vote. If you think the position of the Block Society of the North University of Science and Technology, Blockchain Research Society is consistent with yours, you can entrust it to the Blockchain Research Society of the North University of Science and Technology. Then if you encounter a motion that you want, you can say yourself I don't want matter to help me vote, I want to vote myself. Not yet, right?

High: Now.

Dr. Bao: Direct, real direct democracy, now is mobile democracy in is?

Gao: Not now. My design inside is not, because he can turn at any time. So the difference with his own vote is very little. Because he only needs to turn to one with his position on the same can.

Dr. Bao: Just for five seconds, if I find out that the MATTERS I like, how to think differently from me on this motion, I'll immediately release my commission. I'm putting my vote on another person.

Gao: The only difference, the only difference with direct democracy may be that if all the legislators, the verifier does not have a position with you, then there is really no way.

Dr. Bao: In fact, there may be your problem. I think this mechanism is quite good, we can think about it. In fact, democracy now really encounters many challenges.

Gao: So mobile democracy is forcing that representative to listen every minute of every day.

Dr. Bao: This is in fact the book of the theory of unity of all people has talked about. However, I think we need to see the actual work to know. Because he has been talking about doing a bad job will step down. In praise of republican logic, it is likely to step down in five seconds. Because he can rotate once in five seconds. Then you say to keep listening, this witness or the so-called patriarchs really have a way so focus on the operation of the world and run? In fact, many of these so-called direct democracy, indirect democracy, mobile democracy, we have been constantly thinking about. When this George Washington came down from the Mayflower to establish American democracy, they actually experienced a very huge debate. In their, if I remember correctly, in their democratic charter. In fact, they debated whether everyone should have an equal and equal right to make decisions. In the end, they chose to call those with knowledge and wealth to be the so-called legislators. Right? If I remember correctly, the House of Representatives or whatever house. Many people were discussing, in fact, in a cabin, there must have been a very heated debate. Because wasn't it agreed that all men are created equal? Why should only those who have knowledge, education, and property be allowed to be members of the House? LikeCoin, after moving to the Like chain and then launching the Liker Land app, is actually realizing an imaginary republic and mobile democracy in the blockchain world. Anyone who installs the app can become a citizen of a praiseworthy republic. You buy coins and save them, or you write articles and earn them, and you have a voice in the republic. You can use the credits to vote on motions. Proposals are only allowed at nodes, that is, they are only allowed by validators.

Gao: The current design is like this.

Dr. Bao: But it's actually a complicated system, because for a witness to be a good witness, he has to have a lot of supportive citizens behind him, and that's the next stage. It sounds a bit complicated, but don't forget. From the early days of democracy to the present, we have been going through many challenges and learning. We are still learning. Blockchain could be the next better platform for democracy.

Radical Markets in Taiwan:
Extended Reading with Local Perspectives

Thank you for listening to today's episode, I am Dr. Ruijun Ge. If you like my show, feel free to *click on the subscription button and the next episode will be automatically sent to you. Whether you listen to my show on apple podcast, Spotify, SoundOn, YouTube or other channels, feel free to give me a 5-star rating and leave a comment or a little bell to follow the subscription. See you on the air next time, bye bye.*

* Note: Some of the text in this chapter is marked as time code due to dictation problems, so please go to the original release channel for further information.

11. BBFT PODCAST #EP44 *"THE POST-EPIDEMIC WORLD IS IN THE PALM OF YOUR HAND!"* FT. AUDREY TANG

Originally on SoundOn & YouTube
https://youtu.be/FVA1I4AIOck

Audrey Tang: "Mercury retrograde has just ended, seize the opportunity to bravely place an order with the universe, whether you are like wind, fire, water or land, remember to get the triple coupon!"

Technology, innovation, entertainment, all kinds of novelty, interesting are in Baobao friends say, hey you listen to Baobao friends say?

Dr. Bao: Hello. Welcome to the blog, I'm Dr. Ju-Chun Ko. Today we are not recording in the studio, we are moving all the equipment to the Executive Yuan, the first digital commissar of the Executive Yuan in the physical world. He often says that although it is a digital political commissar, but.

Audrey Tang: There are just more than one.

Dr. Bao: Yes, digital is more than one. We are all part of the digital commissar. There is a concept of I am the swarm. The idol of our information profession is also my idol, and the idol of many people. Ms. Audrey Tang, Political Commissar Tang.

Audrey Tang: Hello PoBo friends said thousands of fans everyone good.

Dr. Bao: Through digital technology, Mr. Tang has solved many problems in the public affairs system for the government and the people. Time is precious today, and it is said that Mr. Tang's time planning is in five-minute increments. We are very fortunate to have more than two units for today's program. We're going to cut to the chase, we usually have a little bit of a long introduction of who the guests are and what their academic qualifications are.

Audrey Tang: My name is Audrey Tang and I have no education.

Dr. Bao: I can't say that. We've heard a lot of great stories that are very exciting and inspiring. We're going to go to the topic that we've been working on today, which is called "From Market Activation to Market Incentives. Let's start with the book, "Marketplace of Motivation". This book is published by Eight Flags Culture, and I think it is the most important book of thought in the beginning of the 21st century. In the process of translating and publishing the Chinese version of this book, Mr. Tang also led us in the spiritual and intellectual aspects, wrote the preface of Radical Markets, and served as the foundation of Radical Markets.

Audrey Tang: RadicalxChange Foundation.

Dr. Bao: It's a global organization. This organization even includes the founder of our blockchain ether.

Audrey Tang: Vitalik Buterin.

Dr. Bao: The content of this book is very widely distributed because of the millions of fans in our audience. Although we have already mentioned the content of the book in about two or three episodes. But could you ask Commissar Audrey Tang to help us increase the number of episodes, and how to make people feel a little bit more interesting about the book?

Audrey Tang: Yes. Traditionally, the West has this idea of the right in the economy. That is to say, it is based on the individual's right to property, the right to capital, the right to run a business. But of course, there is a possibility that this relationship may cause some unfairness. Traditionally, there is a so-called economic leftist idea that not only should there be a social safety net, but also that the government should do redistribution to achieve justice between generations in many matters. The radical market, in fact, is neither left nor right. The radical idea is that we should reimagine a fundamental social structure through the design of a series of mechanisms that will allow people to pursue their own private interests to achieve civil rights (03:23), while at the same time being able to benefit people's livelihood. Or, when we pursue the public interest, we will automatically benefit the person who pursues the public interest. Therefore, through a series of new methods, the individual interest and the public interest no longer seem to be in a zero-sum situation, but in a state of integration.

Dr. Bao: From zero to integration and even to republic, so there are many such discussions. I think I am also very moved. Because often people are used to saying what color you are, white, black. Are you a leftist or a rightist? But in this book, he can say that there is no left and no right. Or super left and super right. I am still digesting this book. The preface was written by Tang Zhengwei, and in a very, very concise number of words, it actually encompasses the wisdom of the youngest economics professor at the University of Chicago, and this little Persian genius of law. I actually want to start today by clearing the air. When I attended some events before, I started to hear people say that Dr. Bao is the first pusher of the radical market idea in the Chinese world. But in fact not, very happy to receive the applause of the political commissar Tang. But.

Audrey Tang: I was communicating with him in English.

Dr. Bao: no more you burst stems. In fact, I myself have always felt that it is not. I will go online to trace the root of a turn. I found in this Tang political commissar long-term accumulation of this call (05:00), not, call records, this conversation records.

Audrey Tang: Verbatim script.

Dr. Bao: The verbatim record website is called archive.tw and is publicly available.

Audrey Tang: Yes, of course.

Dr. Bao: I just found this one from the November 21, 2018, meeting with Grenville. It was a treasure for me at that time. I think it's amazing that in 2018, Audrey Tang was the first one to have the first contact with the beginning of this trend.

Audrey Tang: Yes.

Dr. Bao: Can you share with us a little bit, because at that time the real Chinese world might not have started the discussion. How did you first come across such new and interesting knowledge and the interesting people who brought it to you? What was that meeting like?

Audrey Tang: That one was a video. We had a colleague named Fangrui Zhang (05:48), who was in New York. He was in New York, and he was working with New York** (05:51) to export the Taiwan experience to their Congress. So he met with Glen Weyl, and I kind of possessed him, like that. So it was a meeting of three people. Why did this meeting come about? Because in October of 2018, Glen and I

Radical Markets in Taiwan:
Extended Reading with Local Perspectives

had a mutual friend, Vitalik Buterin, who you just mentioned, and he used to come on Twitter and discuss current events with Glen Weyl from various perspectives. They were just talking about Singapore. And then to say, Singapore in the end is not a you just said this kind of more radical such a place? Vitalik then said you want to discuss Singapore, might as well discuss Taiwan, which Vitalik felt had more of this kind of spirit. Vitalik said, "I know a Taiwanese person, his name is Audrey T. He then tagged me. After he tagged me, Glen immediately sent a private message to say that he would like to have the time to talk to me, in the end Taiwan is, is not how ah and so on. He has been to understand to say that there is a called Henry George thinker later influenced the whole idea of RadicalxChange. So Sun Yat-sen's thought, in fact, especially the part of the people's livelihood. A large part of it came from Henry George's ideas. So he had heard of Su Yi-xian for a long time, but he didn't know many Taiwanese people. So, because of this relationship, we quickly made an appointment, and we made that appointment, just like what you can see in the verbatim transcript. I said to him, you have a lot of ideas in this book are very good. I'm not sure if I'm going to be able to get a good deal on this. But I think we should pick a simple one, so we picked the quadratic voting is square voting method. So, the second year of the Presidential Cup Hackathon, we are now using the quadratic voting method. This is in the "Radical Markets" book, can be said to be the most operational one, because it is very easy. Just raise your hand to vote or post a vote, a little change in the rules can be.

The company's main business is to provide a wide range of products and services to the public. This is also the case with the Squared Vote method, so that you don't have a few projects that are featured on the top of Wired, and everyone is desperate to vote for them. The company also has a number of cold projects, or projects that are actually great but haven't been floated yet, so they can be floated early. Just mentioned this square fundraising method, Dr. Bao's friend said that this program has participated in the fifth and sixth waves of Gitcoin. The sixth wave, because there is no publicity only get eight dollars. But because of this is, there well feel it. That also really feel, because he has a matching, that is, someone cast a dollar. May be because of a numerical calculation process, then there is a pool will give you more than ten to twenty or how much. In fact, it is very interesting. I think listeners, listening to this side of the brain should have been a little warm up. Let's feel the more profound wisdom of Mr. Tang. In fact, the radical market is very grateful to Mr. Tang for his tireless efforts in promotion. In addition to the first type of contact in 2018, the second type of contact was done in books. Then, in the Presidential Hackathon, we also did a lot of pragmatic practices. Then I saw a video later, that is, the publisher should also be here.

Audrey Tang: Yes, the Eight Banners culture.

Dr. Bao: Yes, the Eight Flags culture has come to talk to you. I saw a video recording, and after watching it, I found that the last sentence had a code. This code is a poem.

Audrey Tang: A poem with sixteen pairs of words.

Dr. Bao: You started with the last eight words in the video. It is called the country of citizens, among the flowers.

Audrey Tang: A beautiful island in a beautiful ocean. That is the general history of Taiwan.

Dr. Bao: Can you explain to me is to say that I then looked suddenly a little shocked. Then I went to Google and saw that there was an article explaining this. (10:00)

Audrey Tang: This is our national symbol.

Dr. Bao: Is this our national symbol? The country of citizenship is among the flowers, explain it.

Audrey Tang: Of course I'm speaking with other people on a regular basis in international venues. That includes on CNN, on telecom in many places. People sometimes ask, "What's the name of your country?" I'll say ***** (10:23). It's a cross-cultural country of citizenship.

Dr. Bao: On Republic of Citizens. just now ROC. there is an English one, I would love to hear you read it. Because every time I listen to you read a poem, I feel very.

Audrey Tang: **** (10:43).

Dr. Bao: So I came to read it carefully, and I found that this is actually a kind of Republican complex. That is to say, we feel that we are actually living in the Republic of Citizens.

Audrey Tang: In fact, Korea is also the Republic of Korea.

Dr. Bao: Then why is there this flower? This flower, to me is right up, I found that there is a leakage of some things, this flower is what means?

Audrey Tang: Flower means culture. That is to say, in the long ago, the word flower is originally talking about culture. Like say, we say,

Wanhua. Wanhua is very, very different cultures. So a long, long time ago to now, on this island, of course, there are very many different cultures. We now have more than twenty different national languages, each of which is a variety of different cultures. So it's not that one culture is trying to suppress other cultures in a certain name, only that they are authentic. Rather, our nation of citizenship is built on the recognition and perhaps the revival of various cultural traditions. Then you can call it Zhonghe Wanhua, which happens to be the name of both places. So this is in the hope that when you think of our country's name, it is with a cross-cultural perspective to think. Instead of saying that it is a single culture, that is, the Chinese culture to all other nationalities is to become a second-class citizen. We are all part of the cultural circle of this culture in modern times. If you look at your own culture from the perspective of another culture, you will have a clearer way of thinking (12:13). That is, you will have your own way of thinking. You will be more creative and less like the old days when everything seemed to be aligned with the Chinese culture.

Dr. Bao: And the word flower is so well used. Because on this island of Taiwan, for the sake of civil rights and for the sake of our lives, there have been many flowers blooming in the process of this argument.

Audrey Tang: What lilies ah sunflower ah.

Dr. Bao: It's very, very good. And during this time, especially when we talk about the year 2020, which was a year of dramatic change. Taiwan is not the only country that has birthed these achievements in the past. In this year, in fact, we can kind of say that we can lead the world. We've done a lot, and we're talking about being ahead of the curve? It's all about looking ahead.

Audrey Tang: Over-deployment.

Dr. Bao: Over-deployed, yes. I am really very impressed myself. I first self-admitted that it was the first time that the epidemic happened. Then, our director said to build a factory to carry out this called the national team. I thought I didn't understand it, so I thought, "Is it really necessary? I'm really not sure how much money and manpower it will cost. But now, we can see that we can have graduation ceremonies in Taiwan, we can have all kinds of activities. Every day when we walk out the door, we see a lot of traffic. I think it's really important to be ahead of the curve.

Audrey Tang: More than 90% of people use our physical vaccine, because the biological one has not been developed.

Dr. Bao: Yes. And I learned later that there are a lot of possible risks associated with vaccines. Physical vaccines.

Audrey Tang: Physical vaccines are obviously not a risk for everyone.

Dr. Bao: wear it on the wear right, home can still take it off right. That's why we recently had a new over-deployment, called triple roll.

Audrey Tang: Triple coupon.

Dr. Bao: It's not embarrassing to have a correct pronunciation. We SoundOn, the sound of the program must be correct. Triple coupon. Triple coupon, in fact, I think it is a radical market to stimulate the market. It is not necessarily useful for the time being to stimulate the logic of the market, but his purpose may be a little closer to using market forces to revitalize consumption, revitalize the economy. The company's main goal is to provide the best possible service to its customers.

Audrey Tang: Great. So have you brushed it?

Dr. Bao: Not yet, not yet. Maybe when I just took a cab, I might have.

Audrey Tang: I brushed it yesterday.

Dr. Bao: That also chat, let's say first, we are still actively encouraging people to digital bonding.

Audrey Tang: Swipe three thousand and two thousand, you can withdraw cash from the ATM.

Dr. Bao: I heard that the physical coupons are too beautifully designed again.

Audrey Tang: Too fine.

Dr. Bao: Now the percentage of the collar is said to be a little less.

Audrey Tang: But of course, digital we also have artfun coupons, there are also farming coupons, there are also animal health coupons, there are also what guest house coupons. (15:00) Those will expand the digital part again.

Dr. Bao: Yes. Let me make a call. Really, I'm about 14 days into this, right from the first of July. Today's recording happens to be on the 15th of July. Within these 15 days. My wife has been asking me every day that I should hurry up and tie it. Although I am in the digital field, I am a little worried that I will not have to take out a lot of information.

Audrey Tang: Of course not.

Dr. Bao: To install something.

Radical Markets in Taiwan:
Extended Reading with Local Perspectives

Audrey Tang: How is it possible.

Dr. Bao: I thought yesterday that I can't, I'm going to visit Audrey Tang tomorrow, so it's not like I'm not tied up again. I opened the web page, Googled my credit card company and left a box empty for the vibe triple coupon.

Audrey Tang: Yes. We are a decentralized structure, you don't even have to go to our website.

Dr. Bao: Super go to the center, immediately the first click in, and then enter my ID number, my credit card number. There may be another one that is.

Audrey Tangfeng: Birth date, right?

Dr. Bao: No, not even that. A Captcha.

Audrey Tang: Prove that you are human.

Dr. Bao: Prove that I am human. Then I thanked you that the bonding was done. I was going to record a video, a screen recording to educate my audience. It was too late to start the recording program and it was done.

Audrey Tang: In fact, you cut the picture and the video is the same.

Dr. Bao: There I have a screenshot. Then I thought, "This is not a lie, right? Because he gave me a number. Then I went to that 3000.gov.tw query and it really came up.

Feng Tang: Bound.

Dr. Bao: Can you share with us the whole process of conceptualizing the idea? I know it must be very long. However, the political committee will increase the ego, so it can be shortened. In between, it must involve a lot of data binding, data linkage, and even interdepartmental. The website also allows you to find out where you can spend your money. I immediately went to the places where I usually go to see movies. There are all of them. He will tell you, this theater can be that one can.

Audrey Tang: As long as it is out of the door consumption can be.

Dr. Bao: Just like that credit card I tied to go swipe. I'll just live my normal life. What happens then?

Audrey Tang: you brush you will see you in the time of bonding, he is your bill directly deducted two thousand, or you very much like to go to the ATM then you can also directly take two thousand out.

Radical Markets in Taiwan:
Extended Reading with Local Perspectives

Dr. Bao: First of all, tell us about this wonderful process, was there any conceptual process, any difficulties and good points in the inter-ministerial communication?

Audrey Tang: Yes. In fact, the digital coupon part has not been too difficult. It's a very easy thing to say, because it's easy to swipe 3,000 and get 2,000. One of our biggest controversies in the inter-ministerial time, or do we want everyone to tie in a single website? Or can we authorize each different credit card, ticket, or mobile payment to bundle? Of course, each has its own advantages. If you bundle in a single website, your user experience will be the same. But the downside is that if there is a traffic jam, then everyone will be stuck here. The decentralized system has its own risks, because it's like we can't go through the code of each different platform one by one. So, if there are some specific banks, he said that the first time I swipe the first stick first time card, how about the first number of privileges, this kind of queue, then probably must be the time. The reason is that everyone will flock to the first to win. However, it is only this bank, others will not be affected. Therefore, the user experience for some people is probably not too good. Because at the beginning of the first second, everyone must go for it. But then, you can be assured that even if you have a little bit of this bundle here, you may have another card that you can go bundle to see. That is to say, he has more fault tolerance. that is to say, when there is a problem, relatively will not all come across the problem together. Of course, after spending a lot of money, we all feel that decentralization has a great advantage, that is, our government will not grasp the details of everyone's consumption. We don't know exactly what people have bought with the triple coupons. We only know which merchants people are at. Because we still have to do that, that is, you just said, out of the door to spend the business of verification. How much money did you spend. Add up to three thousand. After that we will not have any other analysis data. That is to say, it is more able to protect the privacy of everyone. And in the process of these different exchanges, he is taking this we call salt hash. This way, that is to say, the identity card number and so on will not be transmitted in our cross these authorities. The bank where you swipe your card originally has what he has. But other parts, including the Ministry of the Interior, the Immigration Department, and so on. In fact, we do not pass our ID numbers to each other. So this can also be said, we refer to the decentralized account book like ethereum such a design, and then use this tokenize some way to ensure that there are very, very many different friends can go together to make a common interest to everyone. The so-called multi party computation, but it's like we want to know how many non-repeating members there are in the OneCard and EasyCard, but not that they pass their ID numbers to each other. It's a MPC process of salting, hashing, and then comparing with each other. We can say that a lot of new construction materials have come from this side of the ether, so that when we build this kind of project, on the one hand, we can protect our personal privacy, but on the other hand, we can make the system so that it will not suffer from that kind of single point of error. (20:00)

Dr. Bao: That's really clear and unambiguous. When I listened to it, I wanted to say, "Well, let me ask the first stupid question again, is it salt and salt bar or encryption?

Audrey Tang: Add salt bar.

Dr. Bao: Is it really with salt?

Audrey Tang: No, of course it's digital salt. That is to say, each of your, say, ATM providers. This should include China Trust, Cathay United, and Taishin. If you want to receive two thousand, you can specify any one of these three ATMs. In fact, you don't have to specify. All you have to do is say I want to get 2,000 at the ATM and you put a credit card that you have tied up and swiped 3,000. Because now there are credit cards what to borrow money at ATMs, cash advances and so on these services. So put your financial card or credit card into any of these three ATMs. It does not need to be these three banks, any bank will do. This way he will go to the check and say, you have already spent three thousand. Then I will give you 2,000 cash directly. This action, he is through each of the different ATMs. He first generated a messy number. This random number is a possible UUID, that is, a random number that does not repeat. Then, he took the consumer's ID number and added this random number. But the calculation of a very long hash value. This hash value, if you do not know this random number, even if you have everyone's ID number, you can not restore to your ID number from this hash value. So that is to say, when they are comparing which people have received and which people have not, your ID number will not be passed on the network. Then we don't know which ATM each person finally chose. All we know is that if you receive $2,000 here, you can't withdraw another $2,000 at another ATM.

Dr. Bao: So the word 嚴 is the word 嚴 in 嚴密?

Audrey Tang: It's SALT.

Dr. Bao: It's really the salt of the salt bar.

Audrey Tang: Yes, he was doing hash when you added an extra string of words.

Dr. Bao: Thank you for really being increased. Then I think the audience should have a feeling of learning something new. And we have these information under the gatekeeper of the political commissar, and many teams together under the common gatekeeper is no problem. I am also curious to say that so when I have a credit card company is directly deducted from the bill, some I may be inserted into the card and then go to collect the money.

Audrey Tang: Many can let you choose.

Dr. Bao: So, comparing my spending to the merchant is not eligible for this positive list, is it by?

Audrey Tang: After your card is swiped, they are called United Card Center, a joint credit card processing center. So then, every two weeks we will update this merchant list. The card is a red, blue, green or whatever recipient of their card. So, as long as there is a list like this in the United Card Center. Every transaction that comes to the United Card Center, he sees that it is an electronic commerce transaction, that is, EC, you have to play the last three codes of that kind. That e-commerce transactions to come, if he is in this white list inside, that count into your three thousand dollars of accrual. If not, you are this buy a do not know the delivery to the house, you simply did not go out. Then such words, of course, do not count into this accrual. But we at 3000.gov are completely unaware of these. This is handled by the United Card Center when he would have carried out such a screening action.

Dr. Bao: This is really, I can feel that this political commissar is very respectful of the audience of our program. So that information is a little bit more difficult than the other interviews.

Audrey Tang: Indeed, indeed.

Dr. Bao: I will go back and listen to it five times. In fact, I am very impressed after listening to it, because this process involves a lot of communication. Let me ask a question for the people first. During the time of the triple voucher implementation and design, did you get less sleep?

Audrey Tang: It's become more. Complicated problems, I have solved in my sleep. So I have to work overtime and sleep until nine hours.

Dr. Bao: Great, that's really, still need to remind everyone to hurry up and tie it. Because I tied the time is July 15. I found out that it is not possible to talk about super-slow deployment, but it is one step behind. Because there are still over-deployment incentives. I don't know if you yourself know three thousand above there is a draw gogoro.

Audrey Tang: Yes, there are smoking gogoro.

Dr. Bao: July 1st to 15th, Ming Zhen nodded his head. Did you get a draw? No. Did you participate? Because I was tied for just over two hours. But don't worry. There are still weekly draws. From the 1st to the 15th, it's a daily drawing.

Audrey Tang: The 16th to October 31 is the weekly delivery.

Dr. Bao: He has a chance for every purchase. Weekly draw or bundle?

Audrey Tang: Binding.

Dr. Bao: Bind a person to draw. So everyone hurry up and deploy it. We're halfway through the over-deployment incentive, and we'll have this bundled offer until October. I'm going to ask one last question that's a little bit more intense. Although I probably have the answer, I'd like to hear your version in person.

Audrey Tang: Because he is beautifully designed. The power of design.

Dr. Bao: How do you know what I'm going to ask. Super Answer. Because what I want to ask is, some news will say that they spent hundreds of millions of dollars and printed a thousand or so million copies. Now it's already 12 million copies are about to run out or something like that, and it will be printed again. (25:00)

Audrey Tang: Yes. It costs a lot inside his final payment, that is, he then goes to the bank to exchange back to the cost of the cash.

Dr. Bao: I want to ask is to Tang political commissar so several political commissar logic, there has not been a kind of internal discussion, do not want to use this opportunity to dramatically, is anyway on the cost on this, the cost of communication

Audrey Tang: The cost of cashing out, the cost of shipping.

Dr. Bao: Should we simply come to a significant stimulation of this share of digital payments in Taiwan, and then use a reward. The carrot plus, can not say stick. The method of radish plus sticker to let people improve a little. Have you ever experienced such a discussion?

Audrey Tang: In fact, the earliest version was purely digital. But at that time, when the Ministry of Economic Affairs proposed such an idea, we also received a voice that was actually the same as when the mask 1.0. Because the mask 1.0 is the time of the real name system, then the Drug Administration is relatively late to discuss the version. At the beginning is to say, then I will use mobile payment. Because of mobile payment, you can also ensure that a person will not double spend, went to the whole family and then ran to Leroux. After going to Leroux and then run to OK, 7 is to wait for them to be smoother and then go. In fact, this case of mobile payment is almost built-in this can not be a person to buy two masks ability. So we had it all planned out at the time. However, the vice president at that time, Vice President Chen Qimai, told us that if you use digital first, only 40% of the people might be able to get the masks in the end. But in public health, unless 70%, 80%, 90% of the people get the masks, the vaccine is useless because the vaccine is useful only if both sides are wearing masks.

Dr. Bao: He is very fast to have the popularity.

Audrey Tang: That's right. If only 40% of the people in a big assembly buy masks because they will pay with their actions. Then the mask will be made in vain. So the point is not how many masks a person has. It is how evenly the masks are placed in each person's hands. Because of this, we ended up going through the pharmacy first. Although the pharmacy queue is really troublesome, the pharmacist is of course also very hard. But at least there is one advantage, that is, anyone can go to the queue. The last 20% of people who really do not have time to queue, we then design 2.0, 3.0, so that they can pre-order to the supermarket. So I think this time the triple coupon, we also heard from around the world is to say, we can certainly say mobile payment only. but the result is to say, about only 40% of people to the end will come to use. The result is that only about 40 percent of people will use it in the end. Of course, this way, from the perspective of revitalizing the economy is also revitalizing to the economy. But from the point of view of fairness to everyone such small vendors will feel that only those who can afford to install this POS machine can be revitalized, more or less a little unfair. So later there is also the design of the paper book. But, of course, the paper book because of the beautiful design is very commemorative value, so that later choose the paper book is more. In fact, I have a suggestion in the internal meeting, no should be tied when you send a beautiful hologram sticker. The hologram sticker can be stuck on your credit card so that you have a sense of prestige when you swipe it. However, later it seems that there is no money for that printing sticker.

Dr. Bao: Entity NFT.

Audrey Tang: Yes, yes, yes.

Dr. Bao: I think all of our listeners have heard of the previous program, and they all understand and respect our program, and they know that we have always had a blockchain series of cattle.

Audrey Tang: **** (28:32).

Dr. Bao: We still have the double flower attack. We'll go back and listen to it later and re-listen. In fact, I was completely convinced. So I'm going to promote it properly, and then I think it's really important that the two go hand in hand. Don't look at us now as if we have a lot of traffic and business has not changed much. But in fact, the economic changes take time. So you really we are now ahead of the deployment to give you this opportunity to revitalize. The triple coupons, triple coupons rush into the ranks of this experience triple coupons. I think this is a very, very important thing. Then again, we have to come more ahead of the deployment. Super ahead of deployment. This is our program or several episodes, this Bao Bo are very worried. The idiom of "worry about the sky" may be used here. You have just said, the country of citizens in the flower. Here comes a puzzle question for me. What do you think the post-epidemic world is in?

Audrey Tang: Among what.

Dr. Bao: Because I would be very worried that we are very calm in Taiwan now. But the number of confirmed cases in the United States is rising. And then our neighboring country, Japan, is also up and down. In many European countries, when we do online reunion at Singularity University, everyone's children are at home.

Audrey Tang: Yes.

Dr. Bao: Because they are from school.

Audrey Tang: You'll be able to see what a lot of people look like at home. (30:00)

Dr. Bao: Yes, yes. Our first reunion will be like this.

Audrey Tang: In fact, even that senior health officials with us in the meeting is so.

Dr. Bao: Yes, yes, yes, that's the feeling. And then, I would actually be a little bit nervous is to say that the first, the next world economy will not happen huge risks, do we have anything to prepare in advance? Money is going to become bigger and smaller? And then we should not go to buy a, I dare to ask, we are not going to buy a gold ah? Or bitcoin? The second one, in fact, is that countries are carrying out long-range forward deployment. We in Taiwan should also be forward deploying remote corporate culture, remote teaching culture, and remote office. This is something that was proposed by political commissar Tang. Is it necessary to triple to mention it? These two. One is how to deploy individuals ahead of time? The second is how to further advance our society as a whole or the government?

Audrey Tang: I think after the epidemic, the world is certainly in this tent. This tent in the middle is to be a thousand miles away. A thousand miles away is through the long-distance approach that I have just mentioned. I still remember the first time I gave a speech with Dr. Bao, it was about 300 of us in the Executive Yuan, all of us were directors of bureaus. Everyone put a tent, cardboard, in front of them.

Dr. Bao: Virtual Reality Cardboard Eye Glass.

Audrey Tang: Yes, yes, yes. We all wear this VR. wear VR after you can experience although we are sitting at the time is the civil service manpower development center in Fuhua. Although sitting there, but a wear after it seems to be to what the desert inside, or is an island above the type and so on. So this actually tells us that we are in this era of the epidemic. In fact, it is the first time that all mankind together to deal with an urgent problem. If it is climate change, the larger your land area, the less you care, because only small land areas will affect the coastline, accounting for a high percentage of the country. However, if the area is too large, it really feels like something that will only be encountered two generations from now. However, in fact, in this pandemic, there is only a time difference of about two months at most between each place and the other place. Now we see some places where the epidemic is very serious, we ourselves will still think of our situation in March, our situation in February. That is to say, the empathy between people has never been so stimulated as it is now. So even if everyone is in the tent, is in the lock down, closed in their own home. But then, I think this kind of global community is not like the old days when each seemed to talk about his own. We can share our Taiwan model together, and we can export our masks to the same factories. We can put our mask map and so on these citizens' innovation, slightly change a few words, Korea is using it, Japan is using it and so on. The company's main goal is to provide a wide range of products and services to the public. In the past, the global village seemed to be a slogan or a tutorial. But now, it is everyone's daily life.

Dr. Bao: Then I'd like to ask a question that is even less knowledgeable. Everyone will say that the economy may change. Although we keep reminding people not to be impulsive to buy stocks all the time, this thing is very high risk. But some people say, we do not invest, we do not speculate, but the money seems to become smaller. This we began to talk about, when the economy changes, what is the value of things?

Audrey Tang: New Taiwan Dollar.

Dr. Bao: The New Taiwan Dollar is not. Just to clarify, because we sometimes accidentally dissed the New Taiwan Dollar before.

Audrey Tang: No, the New Taiwan dollar is compared to other currencies. Many of our nearby economies have a negative GDP, year over the year, but we are still positive. We are not only positive, but we are quite positive.

Dr. Bao: Yes, and after the triple coupon to see if the GDP can be tripled, you can not play this kind of words OK. But it is true that the value of a currency is related to its economic strength and its GDP growth. So with the successful promotion of the triple coupon. We really have a chance to make the New Taiwan Dollar stronger, so welcome to more.

Audrey Tang: Buy the New Taiwan Dollar.

Dr. Bao: It's a good idea to buy into the New Taiwan Dollar.

Audrey Tang: How to buy.

Dr. Bao: It's easy, work more and contribute more. Then this more with our market economy more communication, you will get the new Taiwan dollar, relatively simple. I just talked about this, I just talked about this, I don't know if I can talk about it, I can't talk about it and then cut it off.

Audrey Tang: Speak up.

Dr. Bao: That is, because we just had to take the accompanying gifts to, I invited our interaction department of the University of North Science and Technology very will draw students.

Audrey Tang: Yes, Raspberry Pi.

Dr. Bao: Then, we asked him to draw a picture of Audrey Tang (35:00). Then, it is the version of the CC created by the previous open source adapted.

Audrey Tang: This is a two-color electronic paper.

Dr. Bao: Yes, Commissar Tang is really great. This is three colors. Black, white and red. Then we asked him to paint a picture. And then it was printed, definitely within 300 yuan frame. But I personally think that the political commissar Tang is also a hacker, this citizen hacker. So I wanted to send him a copy of my own recently played very crazy electronic paper artwork. I will make it electronic, on a real and a virtual so.

Audrey Tang: It was great.

Dr. Bao: Once you come to say no no no, this has rules with us.

Audrey Tang: Yes, more than 3,000 to report to the Department of political wind.

Dr. Bao: Yes. Then this electronic paper is almost two thousand, and then this Raspberry Pi is about a thousand to two thousand. So then, just now.

Audrey Tang: I will buy it with cash immediately.

Dr. Bao: So I got the new Taiwan Dollar. In the post-epidemic world.

Audrey Tang: Very good, strong NTD.

Dr. Bao: We can put it on the side, it doesn't matter.

Audrey Tang: accidentally show the password, no.

I'm not sure if I'm going to be able to get a good deal on this. This is for my own questions. I have several episodes of the show have been exploring the so-called post-truth or super-truth, or post-truth. How to say it, because we see this, I think you should know this steam platform founder company called Valve.

Audrey Tang: That's for sure.

Dr. Bao: Valve has a long cool person named Gabe. that Gabe he said in a recent interview with IGN that Matrix is near, or more closer than you think. is closer than you think.

Don Feng: On board the VR will have it.

Dr. Bao: They are now working on post-VR.

Audrey Tang: Post VR, XR, XXR, 3XR.

Dr. Bao: Yes. 3XR. And then, Elon Musk is going to announce his latest R&D results of Neuralink at the end of August. Then I was thinking, this real thing plus what deep learning this super disguise. I really can't imagine how we are going to look at reality in the future. We have apple's glasses. I now think that you should be the physical version of Feng Tang.

Radical Markets in Taiwan:
Extended Reading with Local Perspectives

Audrey Tang: Of course.

Dr. Bao: But I don't know if I'm going to have my brain increased by some device on the side, right? And then I feel your reality.

Ms. Audrey Tang: Yes.

Dr. Bao: And now there is a lot of news, because a lot of this channel a lot. So we feel as if most people feel that the news is true seems to be true. Especially recently because it is very chaotic, the world after the epidemic. A new virus, a new where to come insect. Where and what. What is the definition of reality, or how should we distinguish reality? Can you teach us a secret? It is in the tent of this post-epidemic world. How can we identify reality in the tent?

Audrey Tang: Dialectical truth. It's called tele epistemology, this teleological theory of knowledge. Simply put, this Schopenhauer (38:11) has a very simple way to discern. It is that if something can feel pain, then it is real. Because pain is real. If something cannot feel pain, then it is real. For example, the New Taiwan Dollar does not feel pain. Or, a company or a DAO will not feel pain. Then it is abstract, it is fictional. That is to say, we are using a lot of fictional abstractions to communicate with each other. But the fact is that one may feel the pain, and the other must feel the pain, that is, the subject of me, and the other may also be a subject in the middle. The other person is actually the one who feels the pain he feels. Sometimes it is sublimated into art, sometimes it is simply to tell his pain. So, for example, at the meeting today, we were watching a film in which he solved the dust problem. In the film, there is a crow's nest that can finally be taken out to dry and not be affected by the wind and sand. He can go on planting his land properly. The director said, we are now in the air-conditioned room inside the carpet, we look at these films, he is actually a reenactment, a representation. but just can not forget to say, in fact, watermelon he is not from the ice fruit room inside, from the refrigerator inside out. You see those, that is more virtual. More solid that watermelon, he is in the field, he is actually hot. And if you start to take the watermelon so directly down, he can not quite eat. He still has to be put to let ripen, not only steak will ripen. All in all is to say, in fact, you have to go through that section, that is, he paid his hard work. It's the hard work of the farmer that the grain is hard. This watermelon finally to the top of your plate, he only has this complete meaning, he only is real.(40:00) Otherwise, you see are some reproduction, some image. Some of these creations that are derived from the theme of this. There is nothing wrong with creating, and it can be very enjoyable. But if there is not one that people can experience, originally it is to pay how much toil, to be able to this process of this today, this is not real, it may be hyper real.

Dr. Bao: I went back to listen to this ten times. This everyone knows that SoundOn, or Apple Podcast or Spotify have playback function. You are welcome to play it back as much as you like, the digital world is this good. But Gabe said that he had a very cool interview. He said that they've started working on something that's hacking our perception. He came to a conclusion that the first one he found was that motor sensors, and visual sensors, are relatively easy to create. They were created digitally. But there was one that he found unexpectedly difficult to create, the temperature, the temperature.

Audrey Tang: is just talking about this, the temperature of the watermelon inside the watermelon field.

Dr. Bao: So we're going to start embracing more now.

Audrey Tang: There is temperature. Communication with temperature.

Dr. Bao: Yes, the official website of the triple coupon has this phrase, you have a slogan.

Audrey Tang: good collar, good use, good stimulation, the warmest.

Dr. Bao: Yes. Well, we are really grateful for this, at such a changing point in time. And then, this day to deal with thousands of machines of the Executive Yuan of the administrative members of the two-digit unit of tomato time, to talk to us from the beginning, we are talking about the new wave of thinking in the 21st century, to the triple coupons we hurry to get, and then to the last of this post-epidemic world market changes. I hope everyone has heard what they want to hear. Let me ask the last question first, we may still have two minutes of feeling. In other words, are you a digital political commissar or a youth political commissar?

Audrey Tang: I can't be called a youth after I turn 35.

Dr. Bao: Is that the classification?

Audrey Tang: Our Youth Advisory Board is based on the age of 35. So when I first came to this office, I was still political commissar Cai Yuling and I was his project advisor. When we started giving speeches together, I was just under thirty-five years old at that time. So I was a young man at that time. Yes, now I can only say that I am the chief political commissar of youth participation in the relevant administration, but not a youth political commissar.

Dr. Bao: Okay. So that means that your schooling process was very special because you also went through this period of youth. In such a post-epidemic time, our education is facing many challenges. I know that there are a lot of new educators coming up. A friend of mine named Cindy, who is a high school medical student, is now coming from Taichung to Taipei to do an internship in an AI company.

Audrey Tang: It's great.

Dr. Bao: Is there anything that could be the last because I think our millions of listeners have a lot of ingredients. Is there anything that can be given to the younger ones. For example, maybe you have met yourself when you were in the second year of your country, in such a time of big changes and drastic changes, is there anything you would like to give to encourage people?

Audrey Tang: Yes. I think, I still the same sentence. That is, in fact, the youth is the direction of development of this society. That this vector. Our older generations are the ones who provide the energy. Of course, the vector must add energy and strength in order to move the whole society together. However, we remember this society, in fact, like I did when I was a child, when it was still martial law. I was not exposed to the Internet until I was 12 years old. So I am still considered a digital immigrant, not really a digital native. Of course, one advantage I have is that I can translate and communicate a little bit between the original culture of paper and the digital culture. But my imagination of the direction of the world is probably not as broad as that of the digital progeny. So I think, of course, the future is coming through young people, so we should try our best to support these directions guided by young people. So everything has a gap, and the gap is the entrance to the light.

Dr. Bao: That was really touching. I especially like how you mentioned the stars in one of your interviews. Constellations and astrology look like a pattern, but they are actually three-dimensional.

Audrey Tang: That's right.

Dr. Bao: No need to feel that people say you are a Leo, or whatever constellation is very good. But as long as you have a direction you move forward, that light has an entrance, you move forward to meet.

Audrey Tang: own horoscope.

Dr. Bao: When you meet different stars, you will eventually recognize your own sign.

Audrey Tang: That's right.

Dr. Bao: I think we did the same thing on this episode. We talked about very many things, very many topics. You can follow the direction of the light and listen to it a few times. And then let these small nodes of knowledge, can be in your brain, forming a map of your own a constellation.

Thank you all for listening. Dr. Bao's friend, I am Dr. Ju-Chun Ko. If you like my show, feel free to click on subscribe and the next episode will be delivered to you automatically. Whether you listen to my show on Apple podcast, Spotify, SoundOn, YouTube, or any other platform, you are welcome to give me a 5-star rating. Feel free to give me a five-star rating by leaving a comment or pressing the little bell to follow the subscription, and don't forget to get the triple coupon soon.

Audrey Tang: Digital bonding. Our Mercury retrograde has just ended, so seize the opportunity to bravely place an order with the universe. Whether you are like wind, fire, water, or land, remember to get your triple coupons!

Dr. Bao: Good great thanks to our Audrey Tang Tang political commissar we will see you in the air next time bye-bye.

* Note: Some of the text in this chapter is marked as time code due to dictation problems, so please go to the original release channel for further information.

Radical Markets in Taiwan:
Extended Reading with Local Perspectives

ABOUT THE AUTHOR

Ju Chun Ko now works as Assistant Professor in Department of Interaction Design, National Taipei University of Technology.

In 2012, Ko graduated from the National Taiwan University with a Ph.D. degree in Computer Science. In 2014, he received National Science Council subsidiaries for his postdoctoral research in the Keio Media Design (KMD) graduate school in Japan's Keio University. Later he became the first Taiwanese to study in Singularity University, said to be the "smartest university in the world." In 2015, he co-launched the Luna 360VR Camera campaign on the crowd-funding platform, Indiegogo, raised US$350,000, received investment from HTC and won the Red Dot Design Award (of the world's top 3 design awards) in 2017. After that, he started to get actively involved in the promulgation of the blockchain industry and its relevant schemes, e.g. the smart contract development initiative, while assisting many corporates exploring blockchain crowd-funding, digital assets transaction, blockchain games and other projects.

www.ingramcontent.com/pod-product-compliance
Lightning Source LLC
Chambersburg PA
CBHW060844220526
45466CB00003B/1233

HOW TO USE 360 X4

A concise technique guidebook on how to use the insta 360 X4 for beginners

HOWARD JUSTIN

Table of Contents

CHAPTER ONE 3

   INTRODUCTION 3

   OVERVIEW OF THE INSTA360 X4 6

   UNBOXING AND SETUP 11

   INITIAL SETUP AND CHARGING 13

CHAPTER TWO 21

   GETTING STARTED WITH THE INSTA360 X4 . 21

   CAMERA OVERVIEW AND FEATURES 21

   NAVIGATING THE TOUCHSCREEN INTERFACE 24

   FIRST STEPS TO USING YOUR CAMERA 31

   CONNECTING TO YOUR SMARTPHONE 33

   MASTERING THE CAMERA SETTINGS 40

   ADVANCED CAMERA SETTINGS 52

   ISO 53

   USING PRESETS FOR QUICK ACCESS 59

CHAPTER THREE 64

   CAPTURING STUNNING 360 PHOTOS AND VIDEOS 64

   USING HDR FOR BETTER PHOTOS 68